COMMON & SPOTTED SANDPIPERS

PHIL HOLLAND

Whittles
Publishing

Published by
Whittles Publishing Ltd.,
Dunbeath,
Caithness, KW6 6EG,
Scotland, UK

www.whittlespublishing.com

ISBN 978-184995-361-0

Printed in Malta by Melita Press

CONTENTS

DEDICATION

Derek Yalden and I studied the sandpipers of the Peak District for many years and when sat by stream or reservoir said that after 40 years we should write up a full life story of the common sandpiper and its close relative the spotted sandpiper. Derek's sudden death in early 2013 put a hold on that project, but this book is an attempt to do justice to his work and that of many people who have written up their observations of these birds.

Derek taught zoology at Manchester University and his main passion was the Peak District. His primary specialism was mammals, and he was president of The Mammal Society and author of standard texts on British and Ethiopian mammals. He was also president of the Derbyshire Archaeological Society, and several of the key sites for unravelling the history of animals in Britain are in the Peak District. His books on the history of mammals (Yalden 1999) and birds (Yalden & Albarella 2009) are good reads and full of information.

Because he watched sandpipers mainly during weekends and evenings and never had a grant from anybody to do so, he modestly described his work on sandpipers as a hobby. He usually insisted that authors of publications were listed in alphabetical order, so his name was last though he had done most. His wife Pat is also a zoologist and was involved a great deal. This book is therefore gratefully dedicated to them.

ACKNOWLEDGEMENTS

Thanks are given to all who have written or talked about sandpipers, and particularly those in America whose work on spotted sandpipers was so comprehensive. I especially thank Tom Dougall who has studied sandpipers near Edinburgh, Brian Bates on the Spey, and Wlodec Meissner on studies in Poland. The other fantastic helpers have been people at the British Trust for Ornithology (BTO) and the International Wader Study Group (IWSG). The BTO provides that infrastructure for surveys and ringing which enable an amateur like me to do things. The IWSG provides a network of helpful waderologists around the world to give encouragement and inspiration. Both organisations have friendly conferences enabling enthusiasts to try out ideas on world class ornithologists. In addition, my thanks to the British Museum of Natural History at Tring for their hospitality when I was checking sandpiper skins from around the world.

More than 40 years ago, the foundation of the Peak District study was several years of ringing by Ted Robson, starting in 1968, which showed that there was much to be learned about common sandpipers and their use of streams and reservoirs. Another member of the ringing team, Bill Underwood, gave valuable support throughout.

Most of all, thanks are due to the Peak District farmers and other guardians of the countryside.

1 BREEDING BEHAVIOUR

PREAMBLE

The opinion that common sandpipers are essentially the same as spotted sandpipers was quite pervasive up until the 1970s. A well-used European Atlas (Voous 1960) describing the distribution of birds may serve as an example. That atlas shows the breeding distribution of the common sandpiper spreading all the way round the world. The accompanying text says, 'Although there is no absolute proof that the palearctic and nearctic sandpipers are in fact one species the conspicuous similarity in general behaviour, voice and winter plumage are strongly in favour of that supposition.'

Detailed studies of breeding behaviour have, however, proved this to be wrong. The **common sandpiper** (*Actitis hypoleucos*) appears quite conventional in its breeding behaviour, with a male setting up territory and attracting a female, and the pair raising a brood together. In contrast, the **spotted sandpiper** (*Actitis macularius*) female sets up the territory, gets a male in and lays a clutch for him, and then another male and another clutch, and then another male and another clutch as long as there are enough males and enough time for this 'serial polyandry' to happen. While a few other species have systems where the female lays multiple clutches for a series of males, these occur in open habitats where the female does not defend a traditional territory. Some claim that this spotted sandpiper system with a territorial female is unique. So why should these two superficially similar species (the only members of the genus *Actitis*) breeding in superficially similar habitats behave so differently in the breeding season?

Spotted sandpipers were very thoroughly studied by Lewis Oring and colleagues from the University of North Dakota in the 1970s and 80s. These studies were done as a series of clearly defined projects, usually done to a statistical design on a private island on a large lake in Minnesota. Viewing towers enabled birds to be under surveillance almost continuously. They produced fantastic data (see in bibliography, references starting Oring, Maxson, Reed).

Common sandpipers have been studied mainly as hobby activities, starting also in the 1970s. The breeding study sites are close to major cities and are on farmland, and so are subject to recreational and agricultural impacts. The broken nature of the stream habitats means it is rare for an individual bird to be in view for long (unless it is resting with one leg hidden and some of its identifying coloured rings invisible!).

The British Isles are at the western extremity of the species' breeding range and the Peak District is at the periphery of the British breeders' range. A Scottish study was started in 1993 by Tom Dougall and is less peripheral. More details on the study areas – Little Pelican Island (LPI), Minnesota, USA; the Peak District in England; and the Moorfoot Hills in Scotland – are given in Appendix 1.

This first chapter concentrates on the observed behaviours on the breeding grounds, whereas the population of breeding birds around the world and their population changes (both in local populations and total populations) are left until Chapter 7. This is after we have looked at other studies around the world during the rest of their year, and their food and their enemies. The reasons for differences in breeding tactics between the two species are left to the last chapter.

The most frequently cited references for spotted sandpiper are:

Maxson & Oring 1980 (henceforth abbreviated to M&O)

Oring, Lank & Maxson 1983 (abbreviated to OLM)

Oring, Colwell and Reed 1991 (abbreviated to OCR)

Oring, Reed, Colwell, Lank & Maxson 1991 (abbreviated to ORCLM).

Much of the information about common sandpipers has previously been published in Holland et al 1982a, Holland & Yalden 1994, Dougall et al 2010, but has been herein generally consolidated and amplified with more behavioural data from log books. Two other frequently cited works are –

Nethersole-Thompson & Nethersole-Thompson 1986 (abbreviated to NT2)

Mee 2001 (abbreviated to Mee)

The key works of reference are abbreviated as follows:

BWP = Handbook of the Birds of Europe, the Middle East and North Africa – The Birds of the Western Palearctic (Volume III, 1983, S. Cramp, chief editor) Oxford University Press.

BNA = Birds of North America. Spotted sandpiper sections (Reed, J.M., Oring, L.W. & Gray, E.M. 2013.) This is an online book accessible for a reasonable subscription.

A great benefit of the 21st century is the internet. The contorted descriptions of bird vocalisations using sonograms and comparisons with other supposedly analogous sounds can nowadays be dispensed with by going to the fantastic website xeno-canto. There you can spend hours listening to sandpipers. Thus in this book the vocalisation references are to xeno-canto recording numbers. The internet also has large numbers of photographs from every angle and every part of the world, so in this book photographs are only used where it is felt they will be particularly helpful. Both the common and the spotted sandpiper species are extremely widespread, and you will probably have seen that which lives at some season in your part of the world.

As this book is primarily written for the British bird enthusiast, the following common names are used for species that appear in it several times: lapwing (for northern lapwing, *Vanellus vanellus*), golden plover (for European golden plover, *Pluvialis apricaria*), knot (for red knot, *Calidris canutus*), dunlin (for *Calidris alpina*), redshank (for common redshank, *Tringa totanus*), greenshank (for common greenshank, *Tringa nebularia*), whimbrel (*for Numenius phaeopus*) dipper (for white-throated dipper, *Cinclus cinclus*).

ARRIVING, ESTABLISHING A TERRITORY AND ATTRACTING A MATE

Both the common and the spotted sandpiper are heralds of spring-time in their breeding haunts, and they are often the only longdistance migrant shorebirds in the places that they breed in. Their arrival occurs with the influx of other popular summer visitors – swallows, warblers and cuckoos.

The common sandpiper study sites are occupied in late April. In the period of study when the Scottish site was observed every day (1998–89) and when most birds were identifiable because they were colour-ringed, it was found that 50 per cent of the males were seen by 24 April and 50 per cent of females had arrived by two days later; the complete arrival was spread over four weeks. When the arrival dates of the two individuals making up 48 pairs was plotted, it appeared that while most males mated with a female arriving a day or two later, four males had to wait 10–20 days. But on the other hand some late-arriving males paired with females that had arrived up to ten days earlier. On the first visit each year to our Peak District study area, when there are any colour-ringed birds present, the running average over the years is that 58 per cent of the birds observed were males and that many of the females also present on the first day were their mates from the previous year. On some occasions we have seen only males on the first day, but never only females that day.

The spotted sandpiper site on Little Pelican Island (LPI) in Minnesota was occupied from early May, and in this case the females were first, with 50 per cent in place by around

Figure 1: A common sandpiper which I colour ringed on passage near the south coast of England. The photo was sent by Mike Crutch two weeks later from northern Scotland in classic breeding habitat: pebbly shore with water percolating through. Nearly all the information in this book has relied on the ability to identify birds as individuals, by using coloured rings.

17 May, and males about three days later; again the arrival was spread over four weeks. A typical temperature profile for the continental climate of inland northern America and the maritime climate of the British Isles shows that both species' arrival occurs as daytime average temperatures reach around 10°C.

When they first arrive they may feed in a wide range of places as they have a catholic diet (see Chapter 5) but they will be close to a shoreline that will later provide their chicks with food and shelter. It is at this time of the year that the behavioural differences start to be apparent; while in the common sandpiper the male is the primary territory holder, in the spotted sandpiper that function is taken by the female. However no-one had reported this until individual birds were identifiable with coloured rings, as the sexes are not distinguishable in the field and many of their behavioural displays are ambiguous, performed by both sexes, or multipurpose. The first arrivals seem to mainly feed and recuperate after their long migration, but gradually the intensity of display increases as they pair up and establish their territory. Both species have a general-purpose territory where all activities can be carried out (unlike some shorebirds such as greenshanks, whose nesting and feeding areas are well apart (Nethersole-Thompson 1951)). During this time the activities that take up the sandpipers' energy are defending their space, defending their mate, and feeding. The common sandpiper male does most of the territory and mate defence, while the female feeds to build eggs – whereas in the spotted sandpiper the female does all these things while the male just keeps himself fit and keeps other males away from his part of her territory until his clutch is laid for him.

Having said that males and females look the same, it is appropriate here to say how they are distinguished in detailed breeding studies. When the bird is caught and colour ringed we measure the wing-length and the weight. For both species the females are larger than males (also known from museum collections) so a good proportion are distinguishable on these measures though there is large overlap. Thus on the breeding area a male common sandpiper normally has a wing-length between 107 mm and 114 mm and a mass from 45 to 55 grams, while females have 111–119 mm and 50–60 grams. If a pair is caught, then the female will always be the larger. A female in the egg-producing phase will be extremely heavy and over 60 grams. Any that are ambiguous(112 mm and 52 grams is common!) and stay to breed will eventually make their sex known by their behaviour. For the spotted sandpiper the biometrics are slightly smaller but the same principles apply.

COMMON SANDPIPER TERRITORIAL DISPLAYS

When there are enough birds present that territorial activity is needed to control an area to raise young, the excitement starts. The most common territorial activity is a diddly-dee-diddly-dee in song-flight or on a prominent perch (e.g. xeno-canto XC319861) but this is backed up by wing raises and salutes. When a bird flies to its mate it will often keep its wings raised for a second when it lands, as a form of greeting. But in aggressive encounters, keeping them raised while showing an opponent that it is seriously defending a spot is frequent. In 57 cases where the sex was clearly determined, 47 were the male and 10 were the female; this activity peaks in the first days of May though it continues until the females depart. An extreme form is a single-wing salute as seen in Figure 2 (see p224 of NT2 for another excellent photograph with two other birds responding). In every case that we were certain of the sex this wing salute was done by a male and took place where there was intense competition for breeding space. Figure 3 indicates the time profile of each of these three displays and gives a rough indication of their relative frequency of recording in our notebooks. NT2 expressed the view that a single-wing salute indicated less emotion than raising both wings, but my view is that it is more emotional. Only a few shorebirds do it; it is a very unnatural posture, as wings normally work together even in extreme displaying birds like grouse and birds of paradise; it is done by males already holding a territory and appears to be the last stage of posturing before real fighting.

The under-wing pattern of both species appears to a human eye to be more complex than usual in waders, but what precise attribute impresses another sandpiper is yet to be clarified. Possibly it is just being complete and pristine that matters. A particular series of displays is described in Box 1.

Figure 2: This superb picture of a Common sandpiper (from istock.com) shows the under-wing pattern during a wing salute.

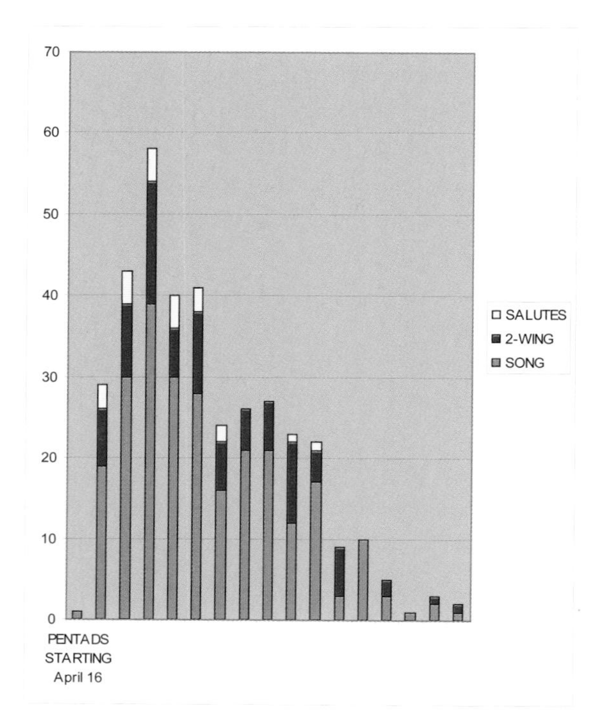

[Left] Figure 3: This histogram shows that display peaks in early May.

[Above] Figure 4: An action shot from Brian Bates of the first common sandpiper that was fitted with a geolocator (see Chapter 2); here fighting an intruder.

Fighting is quite rare. Before it happens there will be some head-down, tail-fanned-out threat postures which may deter the weaker contestant. Once the fight starts it will involve trying to get above the opponent while flaying with wing and feet (Figure 4) and may finish with them both in the water. No physical injury has been reported from these fights between common sandpipers.

A more stubborn but less energetically demanding territory boundary defence may be for two common sandpipers to just stand quite close at the mutually agreed boundary. This is easiest if there is some conspicuous landmark to stand on. This can be an ambiguous sight, though, as occasionally a mated pair may stand together, but generally the female is too busy to waste time standing around.

SPOTTED SANDPIPER TERRITORIAL DISPLAYS

Spotted sandpiper song (e.g. xeno-canto XC232641) is similar to the common sandpiper's song, but is nevertheless distinguishable.

M&O divide spotted sandpiper agonistic displays into five types

- Fights
- chases
- aggressive upright displays
- appeasement displays
- balanced aggressive encounters (parallel walking or facing off at territory boundaries).

BOX 1

ALL THE WORLD'S A STAGE – TWO PAIRS OF COMMON SANDPIPERS SHOW THEIR BOUNDARIES

At 07.30 on 1 May 1989 there were three birds noisy and wing-raising at Chapel Bend in territory 6. I got into a good position to watch. I could see male YB stood saluting and then walking with a two-wing raise to a point where it stood trilling. Then another male OLG with a crouching walk approached, and then they fought in the air about a metre above ground. Female RB was stood on the bank, and after the fight OLG went and stood by her while male YB stood piping on the opposite bank.

OLG raised two wings and the piping got louder. Female RB started to feed downstream in territory 6 towards territory 5; soon male YB went to her and was immediately attacked by OLG. YB then saluted and stood his ground with a one-wing-erect salute, the underneath pointing downstream at OLG. From then on OLG never went further upstream, but if YB moved downstream OLG chased him. Another female, YBW, was further downstream feeding and quietly calling. Then males OLG and YB walked/chased around a pebbly/rushy patch around the territory 6–5 boundary while female RB fed, occasionally raised wings and added her piping to the act. Then it went quiet and YB+RB got on with business in their own territory. The other two apparently disappeared.

The actors in this play:

YB had been ringed as the male in territory 6 in 1986 and had also been there in 1987 and 1988

RB had been his mate from 1986 to 1988 and the pair had some success

OLG had been ringed in territory 5 in 1986, but in 1987 and 1988 had been in territory 3 and in 1989 returned to territory 5

YBW had been ringed as an adult in territory 6 in 1981 and was generally enigmatic:

In 1982 she was only seen once (in territory 7)

In 1983 she was seen twice (30 April and 23 June) in territory 1

In 1985 ditto (28 April and 2 June) ditto

In 1986 she was retrapped on 23 April in territory 16 which is 3.5 km away, but by June was in territory 1 where she bred successfully

In 1987 she was in territory 3, probably with OLG though no success

In 1988 she was successful with OLG in territory 3

In 1989 she was never seen again after the dispute above (nor was she seen in later years)

After YBW disappeared OLG was around until 24 May and then also disappeared for good.

YB continued as the holder of territory 6 until 1995 but in the ten years he was there appeared to succeed in raising only two fledglings.

RB was not seen again after 14 June 1989.

Figure 5: This cartoon of a spotted sandpiper upright display is by Bob Stokes adapted from the sketch in BNA and various photographs of birds.

These are mainly done by females. The difference from what we have seen in our common sandpipers is the upright display sketched in Figure 5 while the single-wing salute is not reported for their spotted sandpipers. More aggressive fights are reported. Two females were noted as being seriously crippled during fights and did not return the next year (OLM). I would guess the development of conspicuous spots and the use of the upright display are related adaptations and may replace single-wing saluting as the final warning.

FIDELITY TO BREEDING AREA

At the simplest level, male common sandpipers are site faithful – around 80 per cent return – and female spotted sandpipers are site faithful with around 65 per cent returning, the smaller number being mainly a result of a shorter life. The female common and the male spotted are less site faithful.

As well as grabbing territory a big advantage of early arrival is that it gives more opportunity for a replacement clutch. In Mee's detailed common sandpiper study there were 1.46 clutches per pair. The advantage of early arrival for experienced spotted sandpipers is the ability to fit in multiple clutches, whether replacements for the first male or for new males. The risk of too-early arrival is that there is inadequate food and a patch of extremely bad weather, so death follows.

It seems unlikely that any mating occurs before the arrival on the breeding grounds as birds do not appear to be in pairs on spring migration and there are no signs of copulation. However they may first meet up in a good feeding area close to their territory rather than precisely within their final territory. What makes a good territory is discussed in Chapter 9.

NEW RECRUITS, ESTABLISHED BREEDERS AND TRANSIENTS

For spotted sandpipers ORCLM gave arrival dates for individuals that had been ringed as chicks at LPI so were of known age and then returned to LPI for multiple years. After the first year their arrival as established breeders was about half a day earlier than the average arrival date of all birds; in their first year they had arrived later than the average, but only two to three days later than their arrival time in later years. For recruits of unknown origin (and unknown age) new birds were arriving over weeks. The bulk of birds arrived between 10 and 29 May but new males continued arriving up to end of June. Females that did not arrive until June rarely got to breed even if they had bred there before. Indeed over nine years of study there were only four experienced females that arrived after 30 May, and they all failed to get a breeding place (Oring & Lank 1982). There were also eight new females that arrived after 30 May, of which only three broke into the breeding population. Males achieved a very different result, however: of 33 occasions that a male turned up after 30 May, only one failed to get a clutch laid for him.

OLM report that the typical composition of breeding females in a year is that 53 per cent are experienced returners, 11 per cent are chicks that had been banded on LPI the previous year, and the rest are new birds coming in from elsewhere unknown and thus may be first-years or have had experience from breeding elsewhere. After nine years of intensive banding of chicks, 40 per cent of the breeding females were ones that had

Figure 6: A spotted sandpiper on show (photo from istock.com).

hatched on LPI. The figure for males was only 19 per cent, indicative that males are the wanderers. By contrast at a sewage lagoon inland, Oring & Knudson 1972 reported that of 27 chicks banded in 1970–71 none had returned by the end of 1972. This suggests that LPI was a much more attractive place for a chick to return to.

There is also a fraction of exploring birds all through the season that are never seen again. In the spotted sandpiper 24 per cent of females and 21 per cent of males that were caught and banded were such transients. This is an area of data where the USA study is much better because of the methodology. On a private island in a lake the water level hardly changes so lots of traps can be left at the water's edge. Every bird that stopped on the island got caught and banded.

For common sandpipers also, a proportion of those ringed are never seen again; in the Peak District our study found 18 per cent, but this is only loosely comparable with the more thorough LPI study. At no time in our study was there a high percentage of the breeding population that had been ringed as chicks in the study area, but we will look further at the issue of recruitment in Chapter 7.

HOW THEY USE THEIR TIME DURING THE FIRST WEEKS

A key question for spotted sandpipers was whether their serial polyandry was leading to a noticeably peculiar use of time by male and female. The private island with special viewing towers enabled continuous observations, and with every individual colour-marked they found it practicable to observe the different behaviours of males and females. A time budget of spotted sandpipers during the pre-laying period is given in Table 1.1 below; the value varied a little through the day and is averaged here (for full details see M&O). For common sandpipers the main question for Yalden 1986b was why shingle is a defended resource when adults hardly use it, and so the time budget was not separated for male and female. This time period between arriving and egg-laying varies from year to year and pair to pair, but is around two weeks.

Table 1.1 *These three categories were not recorded by Y86b; we rarely see nest-building activities in the Peak District; flying and walking were not seen as activities in their own right but merely a means to get from one foraging site or one territorial dispute to another. They were important in M&O because they use a lot of energy.

	Male spotted	Female spotted	common
Foraging	40.8	54. 5	61.3
Preening	26.4	12.1	8.7
Resting	13.9	15.1	14.5
Flying	2.7	1.9	*
Walking	2.9	2.9	*
Nest building	4.1	3.3	*
Courtship	2.1	2.2	4.3
Agonistic	7.1	7.8	11.2

At this stage of the breeding cycle it appears that the behaviours of the two species are broadly similar. There is the need to establish the pair bond and to defend the space to breed, and the female needs to feed to produce eggs. However the male spotted sandpiper spends a very large percentage of his time preening. Whether this is needed to maintain his position as favoured mate for the territory-holding female is not known. It could be a form of display saying to any other male that he is the current choice of this female, but probably it is important due to reasoning described in Colwell 2010: the preen gland produces substances which waterproof and protect feathers and skin. At breeding time, the preen waxes of many shorebirds exhibit a big shift in composition from low molecular weight monoesters to higher molecular weight diesters. It is suggested that this makes the birds more difficult for mammalian predators to smell, thus to change from mono- to diester is good for a ground-nesting bird where nest predation is high. The male spotted sandpiper does nearly all the incubation and this could explain why he spends so much time preening; his life and his clutch depend on it.

Some behaviour is ambiguous. If a bird flies from one end of its territory to the other in response to hearing a call from another bird that could reasonably be lumped with agonistic, but if it is flying to show off to its mate that could be lumped with courtship. Similar comment could be made on walking.

ESTABLISHING BONDS BETWEEN SEXES

We have seen that the sexes arrive at their breeding grounds at essentially the same time, but that in the common sandpiper it is the male that establishes territory and in the spotted sandpiper it is the female. In classic sexual selection the displaying male needs to impress a female such that she becomes his mate. This may take extreme forms in birds of paradise and turkeys. In migratory sandpipers it may be just that a bird has survived the long journeys and arrived back on time in a decent place with enough verve to fly back and forward calling that provides enough evidence. With the common sandpiper this seems to usually be enough for the early arriving females to re-establish themselves with partners from previous years or with neighbours who they recognise from previous years. The balance between rejoining a previous mate and moving to a new mate is summarised below for the Ashop studies:

> On 83 occasions when both of a ringed pair returned they re-mated
> (on 2 of these occasions the pair moved territory but stayed together)
>
> On 22 occasions they split up
> (on 5 of these the male moved territory and in the rest the female moved)

When a female moved it appeared to be a combination of moving to a better territory, or to a more experienced mate, or just that she returned well before her last year's mate and was too impatient to wait for him in their old place. Of the five males that moved, four had only been present one year, and again appeared to have moved to a better territory or to a more experienced female.

It is worth pointing out here that extra-pair copulation (EPC) appears to be fairly high among common sandpipers (see more later) so that a female's 'new' mate may actually be not as new as appears to a casual human observer; the male may in fact have fertilised some of her eggs in previous years.

If the female moves well away from her previous territory this does not appear to be related to any obvious success/failure in the previous year. But again we have to be careful in defining 'success'; the female leaves before the chicks fledge, so she can only see success as chicks surviving the first week, and she is assumed to be intelligent enough to separate the causes of failure at that stage being bad weather, unusually high predation or other facts of life, versus poor male parenting skills.

With the spotted sandpiper, the female has now picked her place but how much choice of mate does she then have? Clearly she can chase away a sickly individual as well as she can chase off another female. Does she just accept that the first males back are likely to be the fittest? As she is going to aim for multiple mates, does she just accept that in order to utilise the strategy of fitting as many mates as possible into the available time the important thing is time rather than quality? If she waits for the perfect male she will probably fail in delivering the most offspring. There was no recorded case of a female spotted sandpiper failing to find a mate. Indeed between 1975 and 1988:

89 territory holders had one mate

65 managed to attract two mates

28 attracted three mates

3 attracted four mates.

Usually these were serviced one after another, but in some cases simultaneously (Oring & Maxson 1978). A bird has been reported (OLM) laying five clutches which included replacements. This sort of behaviour had also been found by Hays 1972 on Great Gull Island in New York, where over two years four females had one mate, four had two, one had three and one had four (see Box 2 'Two years in the life of Red').

Thus the mating choice is little different between the two species – the first birds back are probably the fittest, and they get on with it. The last birds in are less fit and so get less good territories and less fit mates or sometimes no mate.

Ability to regain a territory year after year definitely helps the female spotted sandpiper attract males, as those that only bred there once (43 individuals) only averaged 1.4 mates, and those holding on for just two years (19 individuals) averaged 1.6 mates, while 27 individuals that held on for three or more years averaged 2.2 mates. Looking at it from another angle, a long-lived female that was on LPI for nine years had 17 different males for which she laid eggs (OCR).

Male spotted sandpipers in practice also have on average more than one mate per year. If the male loses the clutch that he is incubating, the female who laid his first clutch may be otherwise engaged with her new male and thus the first male needs to find a free female. Averaged over their study, this meant that males had 1.14 mates per year. This value may be biased a little high, due to a particularly bad year for nest failure in 1975 when 44 of 45 clutches were lost; in that year ten males had two mates, against more typical years when up to just three males needed a different female to replace a lost clutch. Again, looking from another angle, one long-lived male over eight years had clutches laid for him by 12 different females (OCR).

For the common there is a clear task for the male to defend the territory while the female feeds to produce four eggs whose total mass roughly equals her own mass when she arrived. The spotted female appears to have the much more onerous task of defending her territory at the same time as feeding up. The male would at this stage appear to be unemployed but in practice is defending his access to his mate. To do this effectively he has for that period the delegated job of actually defending the territory in partnership with her. There is in reality a similar joint effort for the common, because when there is a serious attempt to steal territory the pair will work together in defence as was seen in Box 1.

BOX 2

TWO YEARS IN THE LIFE OF RED

Helen Hays watched a female spotted sandpiper through two summers on Great Gull Island, New York. She had marked it with a red band (Hays 1972).

In 1970 Red laid a clutch for male-1 in May which hatched on 14 June, and then she laid for male-2 which hatched on 30 June. Male-1 lost the chicks soon after hatching and by then (second half of June) male-2 was incubating, so Red was free to lay a replacement clutch for male-1 which hatched on 15 July.

In 1971 Red was back by 8 May and paired up with male-A. This male arrived on 7 May. He had been on the island the previous year, when he incubated a clutch laid by a different female (Yellow) but only 3 metres away from the 1971 nest. So he still used his patch, and this occurred because Red had expanded her territory to take over space previously occupied by Yellow, who had failed to return. Copulation was seen on 13 and 15 May, and he hatched a clutch on 11 June.

By 20 May she was seen copulating with male-B, and he hatched his clutch on 19 June. On the same day Red chased off her territory an unbanded female who had been showing interest in male-C; both male-A and male-C also joined in this chase.

On 2 and 3 June Red was seen copulating with male-C, a male that had first been spotted on the island on 15 May (when Red was paired with male-A), and he looked after her third clutch which hatched on 26 June. He had been on the island the previous year and nested in 1971, only 1 metre from his 1970 site. His last year's mate, Green, had failed to return.

Red then consorted with male-D which had been around since 8 May, and together they hatched the fourth clutch, on 3 July. This was the only nest where she did incubation. This male had also been paired with Green in 1970. He had been displaying around Red and mate-A the day that they had all arrived (8 May). He seemed attached to a small area around where he had paired with Green the previous year, and so Red with mate-A had moved 150 metres away. In May he was seen with a new female which was caught, and they fed together until 23 May but then she disappeared. Male-D then went and hung around Blue at the opposite end of the island from Red and although an attempt at copulation was seen it was just an attempt. On 4 June male-D chased Blue's second mate, but Blue chased him off and mated with her choice. It was only later on, when Green had definitely failed to return, that Red was seen to join male-D on 5 June; they fed together and displayed, and she laid her fourth clutch for Male-D in 'his' territory; this male was around 400 metres away from the territory that Red had occupied in 1970. In 1971 only clutch 2 was laid in the area covered by her territory in 1970, and a new female, Brown, took over some of her 1970 territory while Red took over territory occupied in 1970 by Yellow and Green.

In contrast to the large area that Red's four nests were dispersed over, Helen Hays pointed out that Blue had three nests in 1971 within 7 metres, and spent nearly all her time close by (and had also done the same in 1970). Male-D needed patience as he was rejected by Red early on and by Blue a bit later, and ended up as fourth choice for Red. How much this was due to his stubborn hold onto a site that was not in either Blue or Red's initial territory is a moot point; maybe he was judged second-rate by the ladies. In Oring's estimation any male eventually finds a female to lay for him, so patience (and even stubbornness) is always rewarded.

Helen watched five spotted sandpiper females on Great Gull Island in 1971. She was mainly there to watch terns, however, so her observations of the sandpipers dropped off as the season progressed. The dates of birds arriving on this island in Long Island Sound (41.12°N, 72.07°W), were from 2 May, so about two weeks earlier than LPI consistent with a maritime climate. Helen still works on terns on the island and sees the sandpipers every year.

When the female is ready for copulation she stands slightly crouched and the male mounts. There is much piping calling and fluttering of the male's raised wings for balance, but the activity takes just a few seconds. It is most frequent in the common sandpiper in the Peak District in the first two weeks of May, and by June the occasional attempts appear to be incomplete. It is not a frequently observed event; our logs average at about one good visual observation per year. That is still better than that reported in BWP for dunlin, where one observer never saw copulation in ten years of watching! Allan Mee, who spent all day every day in 1998 and 1999 watching behaviour, saw 53 attempts. Of these 25 were definitely successful and 16 unsuccessful. The majority (27) were between the social pair, but 8 were with other birds (extra-pair copulation: EPC); 18 were ambiguous as it is not easy seeing colour rings during copulation. Most copulation took place from three days before laying the first egg up until the last egg was laid, but the frequency had been building up from arrival. Copulation by a social pair always took place in their territory, whereas for all but one of the EPCs the female common sandpiper was away in another territory. Storage of sperm is common in shorebirds.

NON-BREEDING BIRDS

Over a seven-year period OLM found 38 occasions of a female failing to establish territory (when 144 successes were scored). Most unsuccessful females left, but six were seen throughout the rest of the season in nearby but poor habitat. This is reminiscent of

Eurasian oystercatcher (*Haematopus ostralegus*) behaviour where there is a waiting area for birds hoping for a territory to become available (Goss-Custard 1996, Chapter 8).

In early years of our common sandpiper study, when there was high occupation of territories there were birds around that we could not pin down to a territory, but with the large shoreline of the reservoirs only a short distance away it would be difficult to say that they were non-breeders rather than being 'on reconnaissance upstream'.

NEST AND EGGS

Both species lay a clutch of four eggs in a nest on the ground. The occasional nest being incubated with a smaller clutch is due to a predator taking an egg or excessive disturbance that prevents the female getting to the nest so she dumps it. Thus Oring and Knudson 1972 report eggs dumped due to fishermen standing by nests, and Walpole-Bond reports one on a railway track and one on a beach possibly due to the egger sitting too close to the nest he was trying to find. Maxson & Oring 1978 showed that clutches with less than four eggs had experienced mouse damage to the missing egg during the laying of the first two eggs when the eggs were unattended overnight; the damaged egg was soon removed by the bird. When mouse trapping was done no further clutch size reductions occurred. Possibly some individuals may lay a smaller clutch due to time or nutrition constraints but this is not proven.

Figure 7: Common sandpiper nest on sloping ground, where the sitting bird has a good view of its territory.

The common sandpiper egg is one of the biggest that there is in proportion to the size of the female, and is possibly the maximum physiologically practicable. A comparison with other birds of similar size is shown in Figure 8, with common sandpiper apparently being the largest of all (26 per cent according to Lack 1966, Table 16). Even the crab plover, which only lays a single egg, is no larger at 25 per cent. The spotted sandpiper is a smaller bird and lays a smaller egg. Whether proportionally it is to same scale depends on what is used as the scale: either a linear measurement (and if so which,

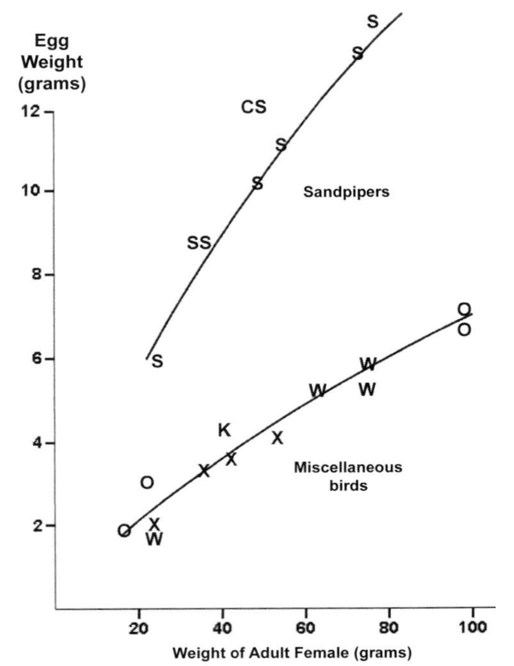

Figure 8: The common sandpiper egg is very large. Data from BWP for familiar birds like thrushes (o), larks (x), woodpeckers (w), kingfisher (k) follow a general line with eggs less than 10 per cent of body weight. Sandpipers (s) eggs are generally greater than 20 per cent, and the common sandpiper egg is very large, at around 26 per cent.

as the birds are slightly different in shape), or mass, in which case a debate arises over choice of a typical weight for the two species.

NT2 describe watching common sandpipers making a scrape, with the male rotating and pressing down into a scoop then the female doing the same, thus suggesting that the site is chosen together. They give a review of nest distances from water, the nature of the site and the vegetation as found by nest enthusiasts over many areas and many years. This and other texts as well as our experience can be summarised thus: nests are most often found around 10 metres from the water but may be several hundred metres away in a forest, or really close, even on a shingle patch in the stream. They will usually be in some rank vegetation, but may be on shingle or short vegetation. The material making the cup is just that local to the nest, and the cup is just enough for the four eggs so is around 10cm in diameter and 3.5cm deep. No great effort is put into nest building, but it will be important that the material below the eggs provides insulation from the cold earth. Sitting on eggs may not sound like hard work but keeping four very large eggs warm is energy demanding.

Eggs are normally laid at roughly daily intervals. Mousley 1939 made a special study of distraction display of birds leaving spotted sandpiper nests. One he found laid its first egg on 15 May and then daily to 18 May. When disturbed on 19 May it left with much display and squealing but the next eight visits it slipped off quietly; then just before hatching it was even more display and squealing. He found this to be a regular pattern; female lays eggs, male nervously starts to incubate and responds to disturbance with intense distraction display, then settles down; but when the eggs are near hatching he becomes hyperactive again. Preston 1951 watched a nest where eggs were laid on 14, 16, 18 and 19 May; so there is flexibility in intervals. The mass of an egg ranges from 8.4 to 10.4 grams, with some individual spotted sandpipers consistently laying large eggs while others consistently laid small eggs; but the layers of large eggs were not bigger birds (Reed & Oring 1997). There were also no patterns within clutches of first eggs being consistently different from fourth eggs, for example, nor did daughters lay eggs similar in size to their mothers.

In looking after the first clutch the next clear difference arises between common and spotted sandpipers. Incubation is almost continuous for the common, with the split of duties being mostly the male doing the longer overnight spell from around 4pm to 7am, and the female the day from 7am to 4pm. The male is thus available to defend the territory much of the day. There is no noise or display at changeover. Indeed during the incubation period the territory is very quiet and, unless diligently sought for, even the non-sitting common sandpiper may not be seen. The sitting bird will normally sit very tight. It seems likely that males are very hungry when they get off in the morning. Mee showed that the time they got up was related to their fat score and was the only response he found from supplementary feeding with mealworms; those males incubated for 1.5 hours longer, giving the female an easier life.

By contrast, the female spotted is often feeding up for a second clutch and defending the territory, and thus plays very little part in incubation of the first clutch unless there is a shortage of males and she fails to attract a second male. Normally the male incubates for around 20 hours.

Loss of the clutch is fairly common in common sandpiper, at around 50 per cent in our long-term study. (This is not based on nests found but on pair behaviour. After a successful pair is seen courting, it then almost disappears and then around three weeks later starts excited displays and chick alarms. A failed pair reappears at less than three weeks without the alarms. Shortly after that it disappears again, then starts chick alarms three weeks after its second disappearance.) The losses vary from year to year depending on weather and habitat. Thus as an example in a study on 10.7 km of the Vistula River flowing through Poland a flood in late May destroyed every nest one year and 42 per cent the next year. Even though this left time for replacement nesting the overall hatch rate was only 32 per cent, but the overall advantage of nesting on islands was shown by their success rate being around 70 per cent while that of those on the mainland was 52 per cent (Elas & Meissner 2016).

For spotted sandpipers it is similar; OLM found 51 per cent of 162 hatched. In one extreme year recorded by M&O, of 45 clutches 30 were lost to mink, 6 to mice, 4 to

BOX 3

AN EGG-LAYING MACHINE

M&O give details of several female histories on LPI. The individual called A:BW laid five clutches for three males starting on 18 May and finishing on 28 June 1976. Each was four eggs. The total weight of eggs that she laid was over four times her own weight.

She returned in 1977 and laid four clutches between 21 May and 27 June. The first male was the same in both years but the third 1976 male was promoted to second in 1977 and a new male was third. The fourth clutch was a replacement for the second mate.

She did not appear to 'blame' a male for losing a clutch; she had needed to lay three clutches for her first mate in 1976 but he was still her first mate in 1977. First come first served appears the norm.

unknown and 4 to floods; just 1 hatched (a 97 per cent loss). That year was 1975, and in the following years some mammal control took place. Hatching was still thereafter very variable. In the best year the loss was only 18 per cent.

In a study at a sewage treatment lagoon in Itasca State Park, Oring and Knudson 1972 reported 50 per cent loss in 1970 (of eight clutches), 56 per cent in 1971 (seven clutches) and 100 per cent in 1972 (four clutches) – at a site surrounded by forest there were at least six bird species and ten mammals seen that were potential predators as well as garter snake (*Thamnophis sirtalis*). This inland site had much less polyandry, and at many nests the female helped incubate when the clutch was complete, because there were no spare males awaiting her eggs. Indeed in 1971 none of the four females were polyandrous.

In summary, hatching success is highly variable due to predation and weather, with 50 per cent being a general level.

INCUBATION PERIOD

In spotted sandpiper the normal period from laying the last egg to hatching is 20 days with some up to a day less. One well-watched nest took 23 days, but 'well-watched' meant the sitting bird was sprayed with paint to identify it and the nest was visited three times a day so it was hardly normal (Burger 1968). In BNA it is reported that incubation period gets shorter as the season progresses, possibly due to warmer weather. For common sandpipers the normal period is 21 days.

Once incubation starts the differences in time budget between the two species become more apparent. The spotted female is still spending most of her time feeding as she builds up for another clutch, but still needs to defend her territory from others. The little incubating which she does is largely in the morning and at the beginning of the incubation period. She has plenty of time to preen. The male is now an incubator and merely spends enough time feeding and preening to keep fit although he engages in an occasional agonistic activity when his clutch is perceived to be at risk or to defend his chance of getting a replacement clutch if he loses the present one. The time spent feeding depends on the food available, but food usually is plentiful and enables a high incubation attendance by the male. OLM report no weight loss for incubating males. Three males were predated (out of 200 sittings). Mark Colwell told me that there is sometimes so much food blowing in off the lake to the vegetation around the nest that the male can feed without leaving its nest!

In the common sandpiper the male defends the territory and forages, while the female incubates during much of the day.

HORMONES

The breeding cycle is accompanied by changes in endocrine gland hormones and these have been studied in spotted sandpipers (Oring & Fivizzani 1991). In males the testosterone levels follow those of normal birds, peaking at his mate's egg-laying stage and then dropping rapidly. Females start with lower testosterone but this increases by about seven times from arrival to the paired stage, and may trigger the territorial behaviour. Her testosterone profile with time is similar to that in polygynous male passerines.

Prolactin is another hormone of relevance to parental behaviour, and male spotted sandpipers had a higher level than females especially during incubation (the opposite from normal birds). An experiment implanting testosterone into incubating males led to a 30 per cent increase in desertion and a 50 per cent reduction in incubation constancy (Oring et al 1989). When the males were implanted with flutamide to reduce testosterone, this led to an increase in brooding at the 1 and 2 egg stage.

POLYANDRY IN COMMON SANDPIPERS?

NT2 describe watching two cock common sandpipers pursuing one hen which then laid two clutches over about a ten-day period in nests about 200 metres apart. She and one male shared incubation of the first nest, which hatched on 1 June, while the other male incubated the second and hatched on 10 June. Derek Yalden had a similar observation in 2006 where one territory had a pair, but in the adjacent territory no female was ever seen. But the male hatched a brood about a week after the pair, and we assume that the neighbouring female gave him the clutch.

Mee reported a female that laid three clutches in total to replace two failures, so the common sandpiper is capable of multiple clutches and so would be capable of polyandry as far as egg production is concerned. Whether the male is routinely capable of lone incubation during British summer weather in our habitat is debatable.

HATCHING SUCCESS

We have seen that about half of clutches get destroyed. If a clutch reaches hatching there is a low percentage of failure of the egg to hatch and most of the failures are physically damaged eggs rather than infertility. Desertion by males also appears to be extremely rare (Mee, who watched every day, never detected it and we have no good evidence for it in the Peak District.)

Figure 9: Newly hatched common sandpiper chicks in the nest.

SEX RATIOS OF CHICKS

Mee, using genetic markers in blood, sexed 124 chicks from 44 broods and found 61 per cent were males. In the 15 broods where all the four chicks were found, 3 broods were all males and only 1 brood was all female. This was just two years in one place (and most of the birds were sampled in just one year) so may be untypical; there was a slight but statistically non-significant trend for more males in early clutches. Andersson et al 2003 also looked at the trend through three years and also found that early broods had a higher proportion of males than par, and late broods a higher proportion of females (65 offspring sampled); they also gave data for spotted sandpipers where no trend was found (129 sampled).

Because the male is the territory holder in common sandpiper, there may be some advantage in early chicks being males and having a slight head start in life. On that basis the spotted sandpiper may be expected to show the reverse tendency but did not.

CHICKS FROM HATCHING TO LEAVING

As eggs begin to hatch, the behaviour of the parent birds changes from unobtrusive to extremely demonstrative. Their repetitive alarms are loud and distinctive (e.g. xeno-canto XC215171 for common and XC31322 for spotted). At this stage it is likely that the response of the sitting adult will be one of their impressive displays of distraction; possibly the 'broken wing' where it struggles along pretending it is unable to fly and dragging a wing along the ground until you get close to it but far away from the nest, at which point it flies off laughing; possibly the 'rodent run' when it runs off squealing in a peculiar crouching style.

Like most chicks they have a tiny 'egg-tooth' on the end of the bill to help them break out (McNicholl 1981). The chicks only remain in the nest for hours (Figure 9) soon leaving and starting feeding. Their legs are strong but in the early days their response to an adult's alarm is to crouch and freeze (Figure 10). However within a few days they know their locality well and the best places to hide and their usual response becomes to flee and hide. The hiding places are under overhangs in eroding banks or in dense vegetation.

Small chicks need regular brooding by adults for warmth; more so in cold and wet weather. In the first few days they probably rarely go for ten minutes without being brooded (see Box 4) and it is assumed that they are brooded all night. Chicks and adults disappear at dusk, and observers use that as a good reason to go to their own beds.

For common sandpipers the data are shown in Figure 11a for weight and 11b for bill for chicks of known age (Holland & Yalden 1991). In Figure 11c mass is plotted against bill length for all chicks caught up to that time, and if we assume that bill length is a good indicator of age it is possible to use all birds caught in a year to get a weight for age that year, then to separate good years for chick growth from poor years. When they hatch, their average weight is 8.7 grams, their bill 9.1 mm; at fledging 40 grams and 21 mm. It is vital that the bill grows to increase their foraging options, so that would be expected to be the most linear. At fledging the wing length averages 93 mm compared with the adult's typical 110 mm. The development of flight capability for migration is covered in the next chapter. The youngest chick observed preening was 12 days.

Figure 10: Common sandpiper chick pretending that it is a pebble.

BOX 4

DOWN BY THE BOARDWALK

Theodora Nelson studied spotted sandpipers at the University of Michigan field station which is on the shore of Douglas Lake. In June 1928 as the first few people were getting ready for the student influx a nest was found right next to the boardwalk connecting all the student houses to the mess hall, with the adult already sat on four eggs. He sat as more and more people arrived, until two weeks later 150 people were walking past six times a day. He got more and more agitated and eventually one egg hatched prematurely. Although it was alive, the small chick and the eggshell were picked up by the incubating bird which flew off to dispose of them. He settled back on three eggs which hatched that night. Next morning the chicks were running about and he led them, as people were going past to breakfast, away from the nest, but stopped to brood them often. By the end of the day they were 140 paces away, where they grew up under a large jagged root on which the father perched as lookout, and beneath which, in wave-created caves in the bank, the chicks fed. As they got older they used around 400 metres of beach, and the father stayed with them for three weeks (Nelson 1930).

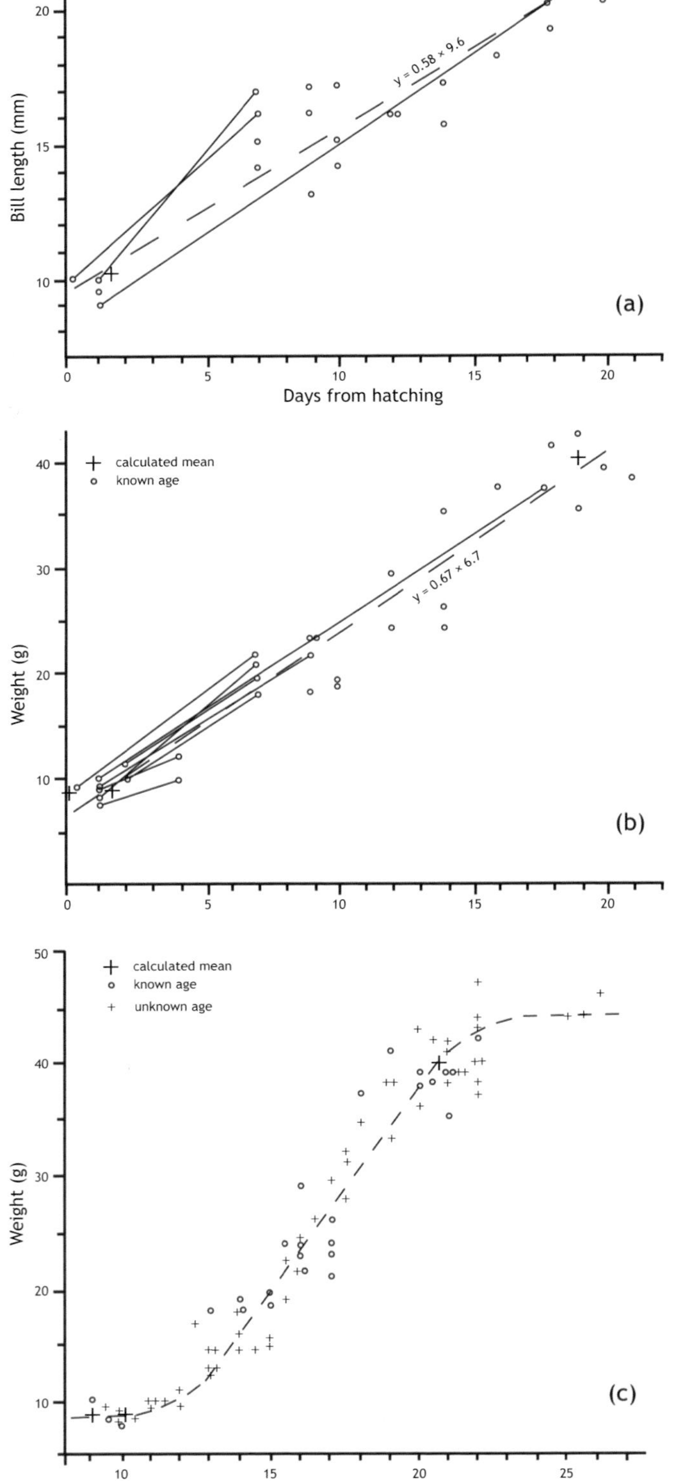

Figure 11: Growth graphs redrawn from Holland & Yalden 1991.

Figure 11a: Growth of Bill

To fit in the egg a chick has a bill less than 10mm long at hatching and it grows linearly enabling the chick to get bigger food items and from deeper cracks day by day

Figure 11b: Growth of body mass

At hatch a chick weighs about 9g. Its growth rate is quite weather dependent Finding chicks that have already been caught once becomes more and more difficult as they get older and it is not good to disturb them from feeding in bad weather.

Figure 11c: Mass versus Bill

Plotting all chicks' weight against bill shows a characteristic shape that has been found for most waders. Assuming bill length will grow steadily because feeding is the primary activity, during the first few days growth is gradually accelerating to a linear growth until flapping starts to use a lot of energy when the rate of increase in mass slows down. Divergence from the average growth curve is taken as indicative of a chick in a healthy/unhealthy growing environment.

The time of fledging depends on the definition. Most are capable of take-off as part of fleeing by 19 days, but before that they will flap as they run to a good hiding place and may occasionally attain lift-off. Over a few days their capability increases and hiding ceases to be any part of their behaviour. At this later stage of their growth they may well be well outside their parents' territory. Indeed, before individually colour-ringed chicks were studied, people quoted fledging times for common sandpipers from 10 to 28 days. Their estimate clearly depended on where the birds observed near a known nest site had actually come from, since a lost brood will generally result in the adult leaving, so then that territory can be used by a neighbour. Observers may also vary their criterion for fledging, from flapping a short distance to full capability.

Nevertheless there is some variability in growth rate. Yalden & Dougall 1994 did a comparison of the rate at which chicks put on weight from place to place and year to year. They compared streams in England, streams in Scotland and reservoirs in England, and concluded that from place to place the weight gain was all generally the same. But comparing year to year there were differences. The best year for weight gain was above average warm, dry and sunny weather, and the worst year was cool and wet (June – mean temperature 14.6°C versus 10.4°C, rain days 6 versus 25, sunshine 206 hrs versus 141 hrs); 1992 was the good year in both Scotland and England, and indeed was also very successful for the number of fledglings produced in the Peak District (Figure 13).

The survival of the common sandpiper chicks is low, at around 25–35 per cent. Many broods disappear totally within days, and a full brood of four reaching fledging is very unusual. Weather appears to be the main problem, though predators are also important. As a result of all the variability in nest hatching and chick survival, the actual fledging success per occupied territory is very erratic (Figure 13).

Figure 12 A common sandpiper chick about seven days old running off after release from the ringer. We only put one coloured ring on a chick. In the first year we put a full code on, but this made the chick conspicuous when it froze, as in Figure 10. If it survives to fledging and returned it will be recaught and given a full code. If it is seen on migration, the white ring will indicate it is a Peak District bird.

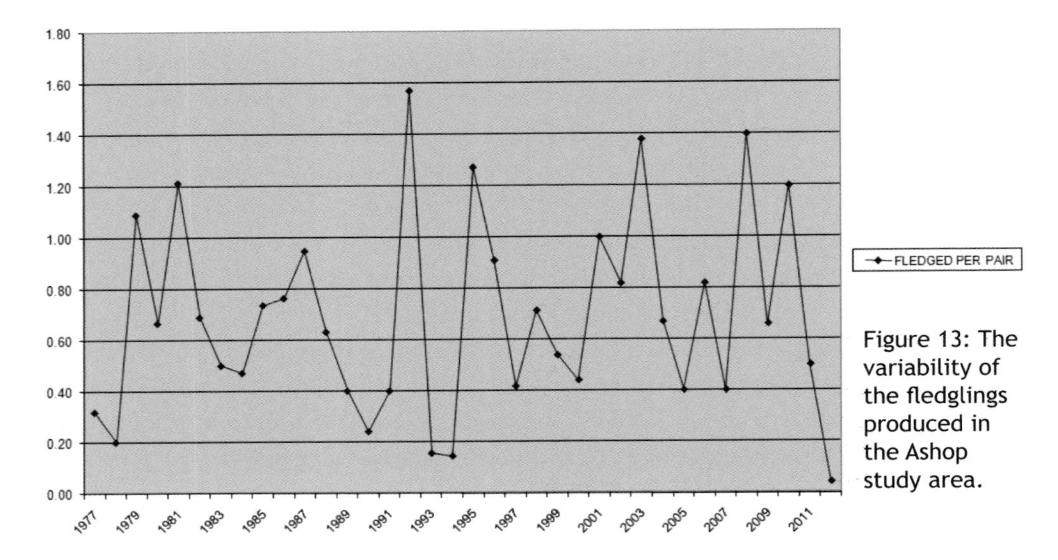

Figure 13: The variability of the fledglings produced in the Ashop study area.

For spotted sandpipers the success from hatching to fledging appears slightly higher, with an average of 42.8 per cent quoted (BNA) though also varying greatly from year to year, with a minimum of 11.3 per cent and a maximum of 81.4 per cent. The quoted fledging time for the spotted sandpiper at 18 days is slightly shorter than the common sandpiper, in line with the spotted being smaller.

HOW THE ADULTS SPEND THEIR TIME DURING CHICK CARE

Most parental care is done by males. In the case of the common sandpiper the female shows steadily less interest from hatching, and is rarely seen right through to fledging. Mee watched at the Scottish sites every day in 1998 and recorded the day the female left. His graph of leaving day versus first egg day has a gradient of close to −1; so the later the brood, the younger the age of the chicks when the female goes. Several females appeared to have even left before hatch; a behaviour not so different from a female spotted sandpiper going off with a new male. The gradient of −1 implies the female feels a strong urge to leave around 20 June come what may. This is consistent with our Peak District observations of date last seen, since half our females had left by 20 June. It is also consistent with the facts that by then:

- successful early birds will have fledged young
- unsuccessful birds will have decided it is too late to start another clutch.

The remaining females are looking after replacement clutches, with a tail-off lasting right through to 20 July. That is because it is just about worth laying again if starting in mid-June so a very late clutch will hatch in early July and a female may sometimes stay till near fledging.

The female spotted is starting to look for a second mate as soon as her first one is incubating. However, if by the time of hatch she is not engaged in egg laying or mate

Figure 14: Spotted sandpiper chick; I suspect that if I found this in a common sandpiper territory I would not notice it was another species. From istock.com.

courting she may be involved with chick care at a similar level to the female common. However, for a typical male spotted sandpiper that is almost solely responsible for a brood, M&O showed that as food increased he spent more time brooding the chicks. This suggests that there is a rate of chick growth that is natural and that getting extra food above that which is necessary is pointless; being warmed, resting and being less likely to be spotted by a predator are more useful ways of chicks spending time than gathering extra food. Nonetheless during daylight hours the male is on average feeding alongside the chicks for 50 per cent of the time, and for the 20 per cent of the time 'resting' he is watchful for predators and 5 per cent of time is still defending his territory. Interestingly some chicks were killed by neighbouring adult sandpipers (OLM) something not reported for common sandpipers and probably a response on LPI to the intensive competition caused by the extremely high population where males were constrained to 20–30 metres of beach.

Another possibility is that the male common sandpiper can be more demonstrative and distract predators and keep the family aware that the danger is still about while the female and the chick actually carry on feeding in a discrete place. If death at the chick stage is the main problem, this would make it worthwhile the female looking after chicks rather than laying more eggs which is of more benefit if egg predation is the bigger issue.

The common sandpiper chicks, once they are mobile enough at about five days old to flee and hide, start to use the shingle to water edge as their main feeding area, while it remains a low-usage area for the adults, who are mainly picking up food near their lookout places up on the banks. It should also be commented that Derek Yalden did not do early morning observations, and the foraging time for adults may be found to be higher earlier in the day (though there was little variation through the day for spotted).

While the male spotted sandpiper is caring for his brood he essentially carves out a part of the female's territory as his family's territory. Eventually there could be four male territories dividing up the female's territory.

Table 1.2 A comparison of time budgets while guarding chicks. In line with the fact that death of chicks is a significant part of breeding loss, the common sandpiper adult now spends most time watching and alarming. The spotted male, who has survived on minimal rations during incubation, spends more time feeding with the chicks. There could be an observer factor here related to whether the bird is more wary due to the presence of his known foe - the ornithologist - and in the Ashop we have been visible, while on LPI they observed the birds from tower hides.

	Male spotted M&O	Male common 15–19 June Yalden 1986b
Foraging	48	9
Preening	12	3
Resting	20	10
Flying	3	*
Walking	6	*
Brooding	6	8.3
Courtship	0	0.7
Agonistic	4.5	
On guard/alarm		68.6

REPRODUCTIVE SUCCESS OF INDIVIDUALS

The factors affecting the year to year reproductive success of male and female spotted sandpipers has been reported in detail (ORCLM). By far the most influential variable is the year. There are good years when more males arrive, and bad years when there are fewer males, and also more predators than average are around.

For individual females the other very significant factor is the number of males they lay for – which, as well as year-to-year variations in male availability, depends on the female's experience. ORCLM show that most first-year females only have one male and raise 1.27 fledglings, but those (30 per cent) that do get more males raise 2.46.

By their second year those only having one male still produce essentially the same at 1.22, but now 56 per cent of females have more males, thus raising 2.57 fledglings per female.

By their third year almost all females are in the same territory they held the previous year and are attracting multiple males and producing 3.27 fledglings. The males are often ones that they have bred with before.

The statistics showed up a small effect of territory size as might be reasonably expected, but for most territories overall quality is not easy for Homo sapiens to judge. In Helen Hays' study on Great Gull Island some female territories are shown as non-continuous, and some overlapping with neighbours and changing to incorporate new bits as more males appear and take up their own space. It is the quality of the bird that is crucial, and her ability to get back on time to reclaim her core territory and appropriate extra areas as opportunity arises.

For successful males there was less variation. If a female judged he was worth the effort of laying a clutch for, and he could protect his bit of territory, he would have a 50 per cent chance of successful hatching and success was mainly driven by predation. If he was successful he might have from one to four fledglings with an average of around three. Statistical tests showed that no benefit was seen as males got older; no benefit was found if the female assisted the male, and no effect was found from variations in measured food availability. Figure 4 of ORCLM suggests that the availability of both birds to look out for the brood is slightly beneficial even if it does not meet statistical significance. This is logical, as a brood of chicks, as they get larger and are more dispersed, will be better defended by two adults. The spotted sandpiper is sometimes described as if she is totally an absent mother (a female version of a ruff as an absent father) but females did some incubation at 65 per cent of nests, and did some care of 34 per cent of broods.

Lifetime reproductive success (integrating over a whole lifetime) was reported by OCR for spotted sandpipers and by Holland & Yalden 1994 for common sandpipers. OCR reported that there is enormous variability, with individual females laying from 4 to 77 eggs while a male may look after 1 to 54 eggs. In both species, the most common category of lifetime reproductive success of males is zero. But total failure of a female spotted sandpiper is much rarer, as she often has more males working on her behalf. Most females produced between 1 and 4 chicks, but only 3 (of 66) produced more than 20. The best male produced 5 offspring that not only fledged but also returned to breed, and the best female produced 8 that returned to join the breeding group.

The most productive male common sandpiper (out of 69) produced 19 fledglings, and the most productive female produced 14 – but very few ever came back to the Ashop to breed.

The benefit of experience is also apparent in common sandpipers. Over two detailed years of study by Mee, old pairs produced 1.12 fledglings while new pairs produced 0.74. In our Ashop study over 30 years, the 37 pairs which were both newly ringed birds raised on average 0.43 fledglings, while those 129 pairs where both birds were known to be returners from previous years raised 0.88. Ultimately the most successful individuals are those who live for a long time, so it is their whole-life decision making that is crucial rather than some detail in their breeding territory.

EXTRA-PAIR COPULATIONS {EPC} AND PARENTAGE {EPP}

When detailed studies are done on genetic make-up of chicks it is found that some chicks are not the offspring of the male that is looking after them. This was a major part of Mee's PhD thesis on common sandpipers, published more widely in Mee et al 2003. During 1999, in 5 out of 26 common sandpiper broods at hatching there was an unrelated chick; in one brood it was one chick, in two it was two chicks (and these were not from just one other partner), and in two broods all four chicks were unrelated to the father looking after them. There was a suggestion that three of the broods had input from the same father.

The incidence of EPP was found to increase later in the season; indeed none of the 12 broods with a first egg date before 16 May had one, while 5 out of 14 afterwards did. Thus it appeared that the early pairs (who were mainly experienced) did not do it. As the season progressed there may be females that had stored sperm from early mating or otherwise felt there was some advantage in seeking out sperm from a better male than her current social mate. More EPP chicks reached fledging than within-pair chicks, but it was a small sample.

Blomqvist et al 2002 also looked at extra-pair parentage in 15 broods (total 53 chicks) and found one chick which was from EPP (ie the female at the nest was fertilised by a different male from her mate) but three chicks were from a different female which had been fertilised by the territory male (called quasi-parasitism). The work also looked at larger numbers of Kentish plovers, and suggested that birds whose social mate was a close relation were more likely to have eggs from mating with others and that this helped to reduce inbreeding.

In 1991 on LPI a number of eggs of spotted sandpipers were taken for analysis, and so the females laid again and again (Oring et al 1992). From this experiment the parentage of each egg was determined. Of 11 eggs that were not those of the female's present mate, 10 were attributable to an earlier mate due to the fact that in birds sperm is stored for many days. In two cases, the males that had fathered the chicks had left the island 16 and 18 days before the eggs were laid! This shows a major benefit in being the early male, as your sperm may still be around even if you have left and another male looks after your chicks. It is still worth the later male doing the tending because if it survives and gets back the next year he will be recognised as a good male. And anyway there were still 66 eggs that had been fertilised by the present male. The 1 of the 11 that did not fall into the 'stored sperm' category was in a first clutch and from an unknown father. There appeared to be fewer cases in this study, compared with Mee's common sandpipers 53 times, of the female spotted sandpiper having EPC with anyone surreptitiously. This is deduced both by genetic analysis as well as the fact that in one study during 1,190 hours of observation of sequential pairings no surreptitious EPCs were seen. Colwell and Oring 1989 reported thoroughly on all the observations of EPC from 1980 to 1985; they showed that it was very variable from year to year, ranging from 88 per cent to 22 per cent of females present in a year doing it, and 40 per cent to 4 per cent of males doing it. Most EPCs were done with next-door neighbours, though for 27 per cent of male excursions they were two or more territories removed. Many of these copulations were done when the female was not laying and may not have led to parentage. What was found, however, was that there was a significant tendency for males that had EPC with a female to later be taken on as a mate, so the interaction may be 'getting to know you'. There was also a suggestion that it was the territory rather than the female that the male found attractive.

LEAVING THE BREEDING AREA

The period between fledging and leaving appears to be very variable. From casual observations of common sandpipers there would appear to be the following factors:

Early versus late broods: an early fledgling usually has better food in its territory, but if it strays it will be chased back by adults in neighbouring territories, so it may get strong quickly and soon depart to avoid the harassment – whereas a late bird may have the area to itself but a dwindling food supply, so it both needs and is more able to remain longer.

Synchrony with neighbours: if several fledglings are of similar age (and the adults have mostly gone) they can share feeding areas and leave together as a little group of juveniles.

Suitability of contiguous downstream territory: smaller streams are generally less good habitat; the chicks may be late to fledge, and the birds can move downstream to better areas whose birds have already departed. Thus they may have left the territory but not the general area.

Thus as the year proceeds the definition of 'leaving' gets increasingly loose, and depends on how far the search is taken by the watcher. But the latest that a juvenile (Figure 15) has been seen on the Ashop was 5 August 2011, though it must be said that our visits do not often continue that late in the season because there are usually no birds. That year there were two very late clutches, hatching around 3 July, so a successful fledgling would not be ready to leave until early August. In other years juveniles ringed on the Ashop have been spotted around the reservoir also up to early August. However this is more properly the subject of the next chapter as the start of the southwards migration.

Hays 1972 reported that of ten fledged spotted sandpipers on Great Gull Island, two were seen for four weeks after hatching, two for up to five weeks, three up to six weeks and three up to eight weeks. In all but one case they were using their natal shoreline, and sometimes defending it.

Figure 15: Common sandpiper in the hand, showing the golden speckling that is so characteristic of juvenile plumage, and the down left on tips of feathers.

2 SOUTHWARDS MIGRATION

START OF THE SOUTHWARDS MIGRATION OF COMMON SANDPIPER

As soon as their contribution to reproduction is done, they leave. Half of the females in the Peak District will have left by 20 June. In one extreme case we had a female that returned to its breeding site of the previous year and was seen in May, then recovered in Morocco on 16 June. We assume that it had failed, so decided to set off back to its non-breeding home.

The males leave once their offspring are able to detect danger and respond adequately without the parental alarms. This is within a few days of their fledging. The juveniles then remain for several more days. The mean date when individual males were last seen (1977–2012) is 1 July. There is no moulting on the breeding territory and no noticeable pre-migratory weight gain. Because throughout their breeding activity they rarely fly more than 10 metres above the ground a bird that flies up to a great height is usually gone for that year.

The first leg of the journey takes them to an area to start to fatten up and to join others. Figure 16 shows the places that Peak District breeders have been seen on southward migration. Most of these were within a few weeks of the end of the breeding season. The typical distance is around 200 km, and this sort of distance is also found in general ringing recoveries a few weeks after breeding.

Before looking at the next step it should be noted that within the British Isles there was an opinion that their initial journey was to follow the river on which they bred down to its mouth. Some may indeed do that but it is generally unlikely. I did a small study (Holland 2009) on the River Lune, which flows southwards, so the birds would be going in the right direction if they did do that. Of 39 colour-ringed breeding birds none were seen at the mouth, though one ringed in 1992 was seen in Lincolnshire in July 1995 and in Nottinghamshire in July 1996. Of 60 birds ringed at the mouth of the river, none were seen on the breeding stretch though one was found breeding on Tom Dougall's Scottish study area 200 km north, and one had bred in Northumberland 160 km north. Both the breeding areas up the Lune and the river mouth were very well-watched places, because the former had a team of people who covered 32.8 km for the Waterways Bird Survey

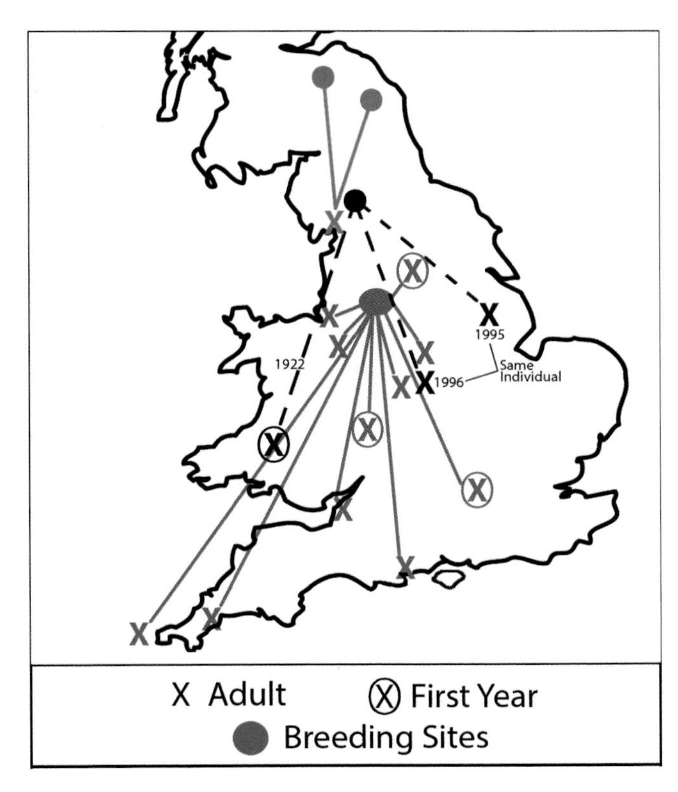

Figure 16: Some stopping places for breeding birds in Britain after they have left the breeding grounds. Two caught fattening up at the mouth of the Lune were from breeding sites 160-200 km north. Birds from the Peak District have been scattered. Birds breeding on the Lune include the very first British recovery in 1922, which was in Wales, and also an individual seen in two consecutive years at different places.

(see Chapter 7) and the latter had a good variety of birds, with a rarity turning up in many years. Other evidence against downriver movement is that most British rivers pass through big towns between the sandpiper breeding haunts, and the river mouth and records of passing sandpipers along rivers through big towns are scarce. Also no bird ringed in our Peak District study was ever reported from near the mouths of the rivers that the streams feed into (Rivers Trent and Mersey). Juveniles often start their journey by drifting a few kilometres downstream.

Of historical interest in British Isles is the very first common sandpiper recovered away from its ringing site, which was ringed as a chick by the River Lune in 1922 and was found a few weeks later 260 km away in Wales.

FATTENING AND GATHERING IN THE BRITISH ISLES FOR THE NEXT STEP

Remembering that a breeding bird (without eggs) weighs around 50 grams and that they quite happily live at 40 grams and can survive at 30 grams, the fact that birds are regularly found above 80 grams at migration stopovers in Britain gives them a capability to get to West Africa without further food.

Figure 17 shows the result of a geolocator on a Scottish bird (Bates et al 2012). It went about 200 km and then remained around for nine days before going another 200 km and stopping for four days (neither of these stays was necessarily confined to a precise locality

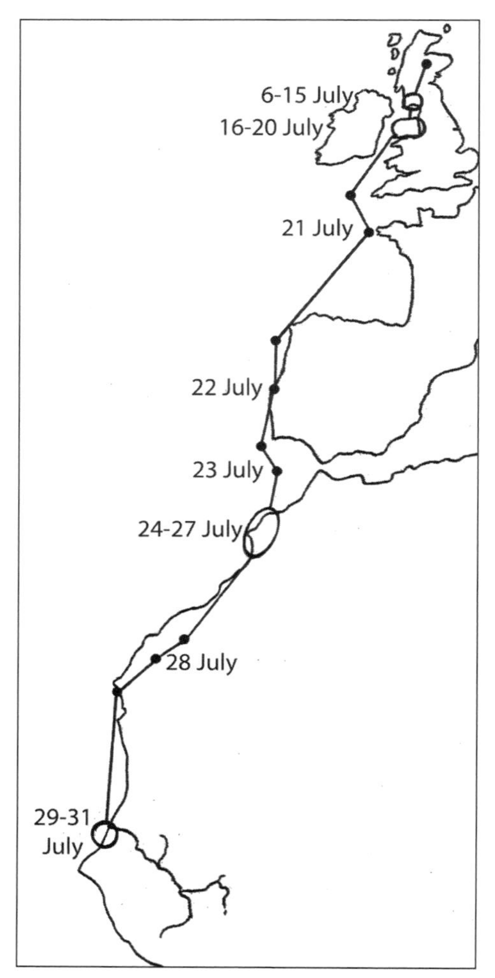

Figure 17: The southwards route of a bird with a geolocator (Bates et al 2012). This one bird gave detail for what had previously been guesses based on the previous 100 years of ringing.

since the geolocators only gave position to around 100km). It then left on 20 July and headed off towards West Africa. It stopped in Morocco for three days. How much food it needed to consume there, or whether it was just a resting place we do not know. We also do not know what this particular individual weighed when it departed. But it is clear from what we will see next that most of its energy requirement was taken on board during its days in central Scotland and north-west England. Its final stage took it to Senegal by 29 July. It stayed around that part of West Africa throughout the winter.

From late June common sandpipers are likely to be seen on any sort of water's edge. The dates of peak numbers differ as we move through the British Isles. Thus for instance at the Lune mouth in north-west England the peak time for adults is mid- July as Scottish and other northern British birds pass through, whereas in the south-east the general (adult and juvenile) median is around mid-August as the mainly Norwegian birds pass through.

Watching colour-ringed birds at the mouth of the River Lune showed that some birds stayed for over a week. If retrapped, they had been putting on weight at around a gram each day. Similar gains were found at Scottish sites (Bates et al 2012) and anywhere they stay on the southbound migration. In July there is plenty of food. At the Lune mouth they feed in the tidal channels where the salt-marsh borders the estuary. Later in the year the estuary will fill up with Arctic waders (dunlin, knot, bar-tailed godwits) but in July, other than the few local breeding redshanks, the sandpipers have little competition. Birds may come back to the same stretch in future years (see Table 2.1 below). The rate of return is around 1 in 6. The birds that were seen in later years had a slightly higher average weight (69.3 grams) than the overall mean. This could be indicative that some of the lightest birds that were caught did not actually survive.

While in their breeding area and in their non-breeding area they are territorial. However, on migration they are mobile and occasionally in gangs of up to ten birds. Thus

Table 2.1 Birds observed that had been ringed as adults in earlier years. The adult birds available to re-encounter each year (Avbl) was derived by assuming that annual survival was 0.8, so that the 5 birds ringed in 1992 gave 4 available in 1993 and 3.2 in 1994 and so on; thus by 2001 for example, there is a fraction of birds available from each of the previous nine years, and rounded to the nearest integer that number is 8. The number actually observed in that year was the bottom row (Obs). The year 2000 is omitted as I was in Canada that summer. A few observations were made by other people as it is a popular birdwatching site. One also observed one bird in 2010 and another in 2013 after I had moved away, but these had been ringed as juveniles (both in 2003).

Year	93	94	95	96	97	98	99	01	02	03	04	05	06
Avbl	4	9	8	7	10	8	11	8	9	11	12	12	9
Obs	1	1	2	1	0	1	1	2	2	0	4	2	1

while they may be seen again in later years, the area they use is not a defined territory, and in salt-marsh channels they may be impossible to see. Thus on 17 July 1996 I colour ringed seven adults but the next day I did not see any of them. I also find it very difficult to estimate how many birds use the salt marshes, as they appear and disappear from view and very few feed or otherwise stop in a place that their colour rings are visible. The chosen small study area had the great advantage of being close to a road and had far better viewing opportunities than is normal in salt marsh.

As well as food, I think that they are looking to join a small flock which is in roughly the same state of preparedness for the next big step in migration. Whereas much of the time they feed in ones and twos (see Figure 18) they may eventually be found in a chattering group of maybe ten birds. They are still feeding, but stick together with contact calls. At dusk they fly off together in a generally southerly direction, gaining height till lost from view. As an example from elsewhere, in looking through county records I found a report of 37 heading off over Rye harbour in Sussex at dusk in mid-August and earlier reports of gatherings on a good feeding area about 1 km north-east of the harbour (Sussex Bird Report 1978).

Figure 18: A superb action picture from Steve Ray of two passage birds at Sandwich Bay in August 2011.

FLIGHT PERFORMANCE AND NAVIGATION

An adult bird has already mastered the air and is able to migrate immediately it is ready, but the juvenile must develop considerably. The distance that an individual can fly then depends on its resources of fat and muscle as well as its choice of route and the timing with respect to weather.

THE DEVELOPMENT OF JUVENILE FLIGHT CAPABILITY

Yalden 2012 described the change in shape of a chick from its first flight attempts, which start with flapping and jumping at about 15 days to what is recognisably flight (fleeing around 20 metres away from danger) at about 19 days with increasing range and confidence thereafter. The shape of a recently fledged individual is noticeably different from an adult (Figure 19). The requirement at fledging is to take off quickly to avoid ground predators like stoats and bird ringers. The shape of the wing is more like the rounded shape of a lapwing, the tail is short and the weight is low. The wing loading is between 0.28 and 0.31 grams per sq. cm (compared with a normal adult of 0.32 to 0.35 grams and a fat migrant of more than 0.5). The juvenile needs to build muscular strength and adequate tail for manoeuvrability, and to grow its wings towards the shape of a long-distance migrant. We thus find juveniles migrating about two weeks after adults. During this period they explore the breeding locality. To survive they need to be able to avoid avian predators like sparrowhawks, which are catching food for their chicks.

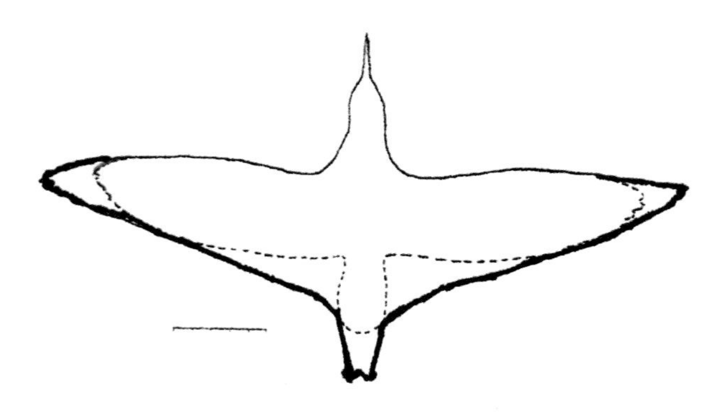

Figure 19: Comparison of a newly fledged with an adult wing shape from Yalden 2012 (the scale line is 50 mm).

MODELS OF FLIGHT CAPABILITY IN STILL AIR

A simple formula for the range in kilometres was proposed by Davidson 1984 for waders, based on their flight speed in kilometres per hour, S, their mass in grams at the start, T, and that at the end, M. He reviewed various other formulae that had been proposed and the evidence of what real birds had been seen to do up to that date. Over time Pennycuick 2008 built a general model of flight, based on aerodynamic principles. According to these models, common sandpipers with the maximum weights recorded are clearly capable of doing around 4,000 km.

ENERGY SOURCES AND BODY CHANGES

Most of the energy is stored as fat, though the birds are capable of using protein as well. The great migrations of knots have been studied very intensely, and it has been shown that there are other changes to organs (to reduce mass of those not essential during migration) and use of other sources of energy (Piersma & van Gils 2011) which probably apply, at least partially, in sandpiper migration. As they put on mass their pectoral muscles get bigger, and as they lose mass the muscles get smaller. In ideal conditions they are capable of putting on weight much faster than the generally observed 1–2 grams per day. A test on six birds caught on southbound migration in Sweden and kept captive with unlimited and easily obtained mealworms showed that they added 5–6 grams per day, which as a percentage of their weight was greater than any of the other 13 wader species treated similarly. In terms of energy assimilation, the rate was 9 times their base metabolic rate and was only exceeded by knot at 11 times (Figure 43 in Piersma & van Gils 2011). Many common sandpipers preparing for migration are caught at over 80 grams, at which weight their fat reserves are at about 100 per cent and they are capable of travelling 4,000 km.

WEATHER SYSTEMS

Studies at migration hotspots the world over have detected the significance of weather, whether it be thermals aiding raptors over Hawk Mountain or arrivals of passerines on Fair Isle. The meteorological side is well described in Elkins 1988. For the southwards migration of common sandpiper from western Europe, a key feature is the prevalence of weather systems in summer giving favourable winds (Figure 49 in Elkins 1988).

REAL LIFE

Some of the wader migrations that have been shown over the last few decades with satellite tracking or geolocators have been surprising and probably 100 years ago would have been thought incredible, like bar-tailed godwits (*Limosa lapponica*) flying direct from Alaska to New Zealand across the Pacific taking over a week, white-rumped sandpipers (*Calidris fuscicollis*) flying from Canada to South America direct across the Atlantic, knots from Australia to China, red-necked phalaropes (*Phalaropus lobatus*) from Scotland to Peru …

Compared with this company we will see that common sandpipers are fairly unadventurous, and mainly migrate over land.

FINDING THE WAY

The juveniles cannot follow their parents, because they have already left. Spotted sandpipers in captivity (Oring et al 1997) showed migratory restlessness when the weather was good. Since all the birds breed well north of the tropics, and most winter south into the tropics –nearly due south – the initial task of setting off in generally the right direction is presumably innate. They may do the first part of the journey alone but soon get into small groups, and except for very late broods will probably meet up with some adults too so that group wisdom gradually increases. They are night migrants, and like other birds will use the stars and possibly a magnetic compass too.

Waders watched on radar have been generally reckoned to fly at 4,500–6,000 metres above ground (e.g. Able 1999). At 6,000 metres the view ahead is potentially about 300 km. The geolocator bird flew (Figure 17) with no time for significant refuelling from north-west England to Morocco, and did that on nearly the direct great circle trans-ocean route to the north-west tip of France then the north-west tip of Spain, down the Portuguese coast and across to Morocco, but was always essentially within sight of land. However the average speed for the straight journey would only be 20 km per hour which is well below the general speed, 45 km per hour, of a shorebird that size so it probably had a break in Brittany (21 July) and Portugal (22 July) during the day where the noon locations are. The journey across the Sahara may also include a rest in the shade on 28 July.

It would be reasonable to assume that a visual map of coasts and mountains and plains gets into their brain. This map will include the major barriers of the deserts, seas and mountains they encounter and will need to cross again or avoid in their northwards migration.

Although they are able to survive long sea crossings, as witnessed by vagrant spotted sandpipers in Europe and Tristan da Cunha and by common sandpipers in Mauritius, they do not appear to choose long ocean crossings as a routine strategy.

ACROSS MAINLAND EUROPE AND SOUTH-EAST ENGLAND AND MOROCCO

There has been a great tradition of ringing and observing passage waders across Europe. The very large weights found in Britain (and in southern Africa for birds preparing for the northwards migration) have not been found to be as prevalent, and thus the deduction is that the strategy of most birds as they cross mainland Europe is to use shorter stages. The places mentioned in the next sections are shown on the map in Figure 20. This is not a comprehensive list of every study ever written, but ones chosen to illustrate some points.

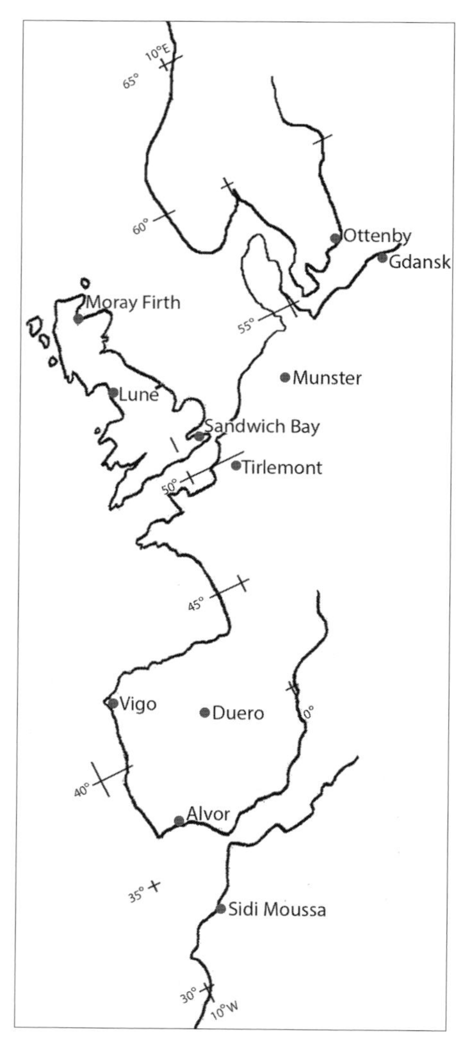

Figure 20: Some key sites where common sandpipers have been studied on their southwards migration.

SWEDEN {OTTENBY}

The absence of significant tides in the Baltic means that a hundred traps can be left out as long as there are people to extract the birds that have been trapped. Iwajomo & Hedenstrom 2011 describes results from 1947 to 2008 during which period 5,232 common sandpipers were ringed. This is a fantastic sample, though it averages just one bird per catching day. Key data are in Table 2.2 below, where comparisons with other sites are seen. The major benefits of this long-term study are the following trends that were observed.

Over the 61 years no significant trend was seen in total birds captured, and in high catch years there were both more adults and more juveniles. However over the years there was a slight increase in the number of adults, and a slight decrease in juveniles. An attempt was made to see if high numbers of juveniles were followed by increased adult numbers, but no such correlation was found.

When the weight of all the birds were plotted against their day of capture, the straight line fit for adult extra weight against date shows a decline; thus on average their early birds are about 5 grams heavier than late birds. However, we know that females migrate earlier than males and that females are bigger than males, so a reasonable allowance for this leads to the conclusion that the average extra percentage weight in the middle of the migration season is a reasonable percentage for any adult (28.8 per cent above a lean weight, which they assumed to be 39.6 grams). In contrast, for juveniles the straight line fit shows an increase. The reasonable explanation of this is the youngsters are still growing generally. However it could be that earlier broods have a higher fraction of males (see Chapter 1 on brood sex ratios). The average weights, around 51 grams, are essentially the same as those in breeding areas where they are also not 'lean'. So at Ottenby they do not have high fat loads.

The next trend of importance is the median passage date, which for adults showed no significant change over the 61 years. However juveniles indicated an earlier median passage. A possibility is that it is becoming more difficult for a very late brood of chicks to reach fully grown, so that is why there are both fewer juveniles and an earlier median passage. Although observation of migration was essentially finished in the first week of September (Figure 21), one late juvenile was caught on 18 October.

POLAND {GDANSK AREA}

This southern side of the Baltic has been a great site for studying wader migration for many decades. Results for common sandpiper from three sites about 20 km north of the city of Gdansk were described by Meissner 1996 and 1997, but he also had access to data from the mouth of the Vistula near the centre of Gdansk.

In Table 2.2 we can see that the dates and weights are similar to those at Ottenby. Timing is given in a different way, in their reports (and based on counts rather than catches) with an emphasis on differences from year to year and indeed small differences between a river mouth site and a seaside peninsular site, so the value given is my overall estimate from their published data. With regard to year-to-year differences, there were two notable extremes, the earliest being in 1988 when 311 birds came through with a

Table 2.2 Dates and weights of birds trapped for ringing at various migration sites (the citation for each place is in the text). At most sites there are birds around 80 grams capable of getting to Africa south of the Sahara with no more food. The average weight of those caught in Scotland and north-west England shows that most birds are heading towards that weight. At good feeding sites in Europe slightly smaller average weights are found, and at places that are mainly resting stops there are few heavy birds. In Spain many birds are going no further, so the average is pulled down to that close to those who remain.

Place + co-ordinates	Years of study	Age	Sample size	Median date	Mean weight/g	Maximum weight/g
Sweden – Ottenby 56°12'N 16°24'E	1947–2008	adult	3,290	21 July	51.0	81.3
"		*juvenile*	*1,942*	*7 August*	*49.5*	*76.9*
Poland – Gdansk 54°15'N 18°30'E	1983–1990	adult	212	24 July	47.8	67
"		juvenile	681	9 August	49.7	71
Germany – Frondenberg 51°28'N, 7°45'E	1951–1960	adult	44	28 July	57.3	80
"		*juvenile*	*60*	*19 Aug*	*60.3*	*80*
NW Spain coast 42°20'N 8°40'W	1991,1993 – 1996		111	Not given	51	65
Spain inland 41°35'N 4°45'W	1986–1994	adult	206	4 Aug	53	73
"		*juvenile*	162	28 Aug	50	69
N Spain	2007–2013	adult	113		51.2	
"	"	*juvenile*	158		46.2	
Scotland – Moray Firth 57°65'N 3°62'W	1970–1982	adult	22	21 July	65.5	81
"		*juvenile*	*107*	*25 July*	*56.9*	*75*
NW England –Lune mouth 54°05'N 2°45'W	1992–2004	adult	41	16 July	67.7	84
"		*juvenile*	*10*	*21 July*	*63.8*	*81*
England – Sandwich Bay 51°15'N 1°25'E	1971–2012	combined	285	Early Aug	60	78
Thailand	2000–2011	combined	41	Aug/Sept	44.2	58
Morocco	1971–1973	combined	34	Aug/Sept	43.1	

median date of 2 August, and the latest being 1989 when only 95 birds were counted with a median on 24 August. Looking for clues in the Ottenby dataset: 1988 was also their largest catch of adults while 1989 had less than half the adults but more juveniles; the overall ratio of juveniles to adults, taken at face value, would imply that each pair raised six juveniles, so clearly the site is not a regular adult stopover. Another factor is that their catching season only started on 19 July, so many adult females and failed males will have passed through before. There is a very low numbers of retraps (Meissner 1996).

GERMANY {FRONDENBERG, GÖTTINGEN} AND SWITZERLAND

Mester 1966 caught birds at Frondenberg in Germany (about 50 km south of Munster), and it is immediately apparent that the role of this site is different from the sites on the Baltic coast. The median weight is almost 10 grams higher, and there were birds staying for up to three weeks between capture and recapture, showing weight gains of 16 grams in 15 days and 20 grams in 21 days, though there were also ones that lost weight of 6 grams over 11 days and another 4 grams over the next 12 days. These latter observations are very odd, as if the feeding was poor for those birds I would have expected them to move on. Maybe they were sick. The attraction of the site is also shown by retraps in later years (three after one year, two after two years and one after five years). The capture timing is only given in ten-day intervals and starting on 20 July, so it is not possible to be sure of median dates, though the first returning bird seen ranged from 1 July in 1959 to 18 July in 1957 with an average of 8 July over the eight years. Thus a median of around 28 July is reasonably deduced, because by late August the numbers of adults dropped to essentially none. The peak of juveniles caught was around 16–20 August, but the odd bird was recorded late, the last being 26 October. Birds staying so they can put on weight will obviously put the median dates for those places later than those of the places where they just pause briefly.

At the settling and treatment lagoons at a sugar factory at Norten Hardenberg near Göttingen, Riedel 1977 caught 616 birds. The main subject being investigated was the time they stayed and whether they returned the next year. He deduced that a typical period for adults staying was 12.7 days (based on observations of 84 colour-ringed individuals) and for juveniles 9.4 days (44 individuals). The return rate in later years was as follows:

Table 2.3, At a good feeding site near Gottingen there were retraps in later years suggesting some fidelity to a stopover especially for adults, from Riedel 1977

Adults: of the	**260**	**individuals that could have been seen**	**1**	**year(s) later**	**26**	**were actually seen**
of the	182	"	2	"	18	"
of the	91	"	3	"	8	"
Juveniles: of the	178	"	1	"	9	"
of the	144	"	2	"	4	"
of the	74	"	3	"	1	"

Birds retrapped in later years as they passed through eastern Switzerland (Glutz et al 1977), showed that four ringed as juveniles came through four weeks earlier as adults, but those ringed as adults came through on average on the same day, and one bird that was caught five years running was on 22, 17, 19, 19 and 31 July.

These data show loyalty to a passage site if it provides good feeding. The recoveries from all of these sites show a general south-westerly movement of the large Scandinavian population heading for West Africa.

BELGIUM {TIRLEMONT}

Vansteenwegen 1978 did a thorough study in 1976 at a sewage farm of around 30 ha: 60 individuals were ringed and 9 were retrapped, and birds were counted every day from 20 July to 10 October. The retrapped birds showed weight gains and weight losses similar to those at Frondenberg. The mean length of stay was assessed as 18.7 days. and the number of individuals that used the site in that year was assessed as 113. There was an attempt to detect what effect weather had on arrivals but nothing was found. It was noted that although they feed alone they do gather at dusk, and it was also judged that the number of birds was not limited by food availability.

SOUTH-EAST ENGLAND

Recoveries shown in the British Migration Atlas show that most birds seen in south-east England are part of the Scandinavian breeding population; mainly from Norway.

Nisbet 1957 wrote up results of 30 years of data collected at Cambridge sewage farms where the peak of migration was 8–11 August; the median deduced from his tabulation was 14 August (6,000 observations) and the last was on 22 October. This puts the south-eastern corner of England on the same timetable as adjacent parts of mainland Europe, as opposed to the timetable of most British breeders, who by early August are already in Africa, south of the Sahara.

Brown 1973 summarised early data from Britain. The maximum weight gain was 3.2 grams per day (but over a short period of four days) while ones that were retrapped after longer periods were more typically 1 gram per day, and one adult averaged 1.5 grams per day over 17 days. There was one bird that weighed the same after a week, but none that had gone down in weight like those in Belgium and Germany.

In Riddiford & Findley 1981, histograms of numbers seen at observatories around the British coast are given. The largest numbers and the longest season are at Sandwich Bay, which is on the south-eastern tip of the country; there the median date of all observations was around 20 August. The ringing data provided by Sandwich Bay Bird Observatory (to 2012) give a peak in early August, and the weights were similar to fattening sites in Germany (but less than in north-west Britain). Catching is most profitable at the peak, and the long period of migration means that the median of visual passage is later than the median of catching.

SPAIN AND PORTUGAL

On the northern coast in the corner of the Bay of Biscay in coastal marshes between 2007 and 2013, it was found that some birds stayed for up to 36 days and the mean stay was

Figure 21

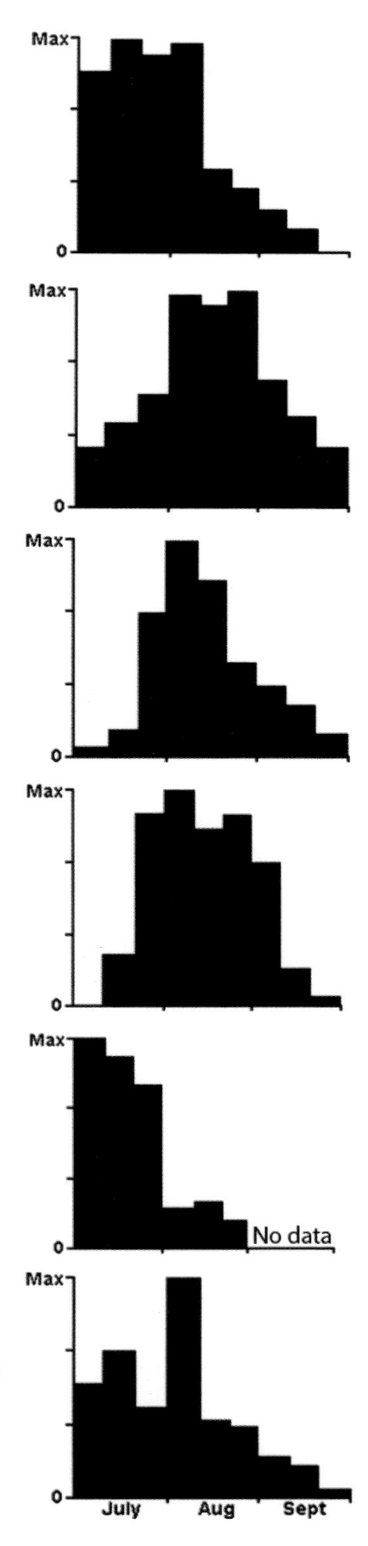

LUNE SALTMARSH

(NW England)

A high number during July from northern Britain many of which stay to put on weight

EYEBROOK RESERVOIR

(Middle England)

Low numbers all summer mainly brief stops of birds from a wide ranging area and more at end of September than early July

SANDWICH BAY

(SE tip of England)

High numbers peaking in August as Norway birds pass through and it has a similar pattern to mainland northern Europe

DUERO RIVER CONFLUENCE

(Central Spain)

Few until late July but a long period of high numbers through August into September as birds from large parts of Western Europe pass

LAKE BAIKAL (N bank)

(Central Russia)

A similar pattern to NW England so guess getting birds from fairly local area north

LAKE ERIE (N bank)

(Fairly middle of N America)

Spotted Sandpipers show a similar pattern to NW England and Central Russia though a massive peak of juveniles in the first week of August masks it estimated as 19 days (de Elgea & Arizaga 2016). Some individuals put on around 25 grams over 10 days while others hardly changed over 25 days. Those putting on over 10 grams were mostly around during early August and were presumably keen to push on to Africa. They pointed out that a study 100 km further west had found that birds stopped for less than a day.

Arcas 2001 did a study covering August to October for five years in a marsh near the mouth of the Lagares River near Vigo in north-west Spain. As well as weighing the birds, he scored the amount of fat on a scale of 0 to 5 which is regularly used for passerines where score 1 is slight fat (on the bottom of the interclavicular region) up to score 5 when fat overflows that region. Nearly all his birds were assessed as 0 or 1. Furthermore there were more retraps than at other migration sites, and in those retraps no weight increases were found. My interpretation of these observations is that most of these birds were going to remain for the winter on this far west coast of Iberia. This is because most British birds will normally have passed over well before mid-August, and the birds from Scandinavia that have passed through Ottenby, Gdansk and Germany then recovered in Iberia have not usually been found that far west. Iberia does have a large population of wintering common sandpipers.

Balmori 2003 & 2005 caught birds at the confluence of the Rio Duero and Rio Cega in the middle of Spain about 140 km north west of Madrid (the Duero in Spain is called the Douro in Portugal and reaches the Atlantic in Oporto). He caught birds at each end of the day and had many retraps including some caught again in 12 hours. He quoted weight gains during the day averaging 6.87 grams, and weight losses during the night averaging 4.94 grams. His longer-term weight gains averaged 1.93 grams per day, similar to other places. The peak weights at this site are above 70 grams, so many birds are preparing for a long journey. This site appears to be similar to many river confluences throughout Iberia, and several of the Scottish breeding birds fitted with geolocators have stopped in the middle of Spain (Brian Bates' talk at BTO 2015 conference).

Data collected in the Alvor estuary in the Algarve region of Portugal by A. Rocha, which were displayed as a poster at the Wader Study Group held in Portugal in 1995, showed birds that were of lowish weight and might again be going to stay around the coast all winter. Indeed the general picture appears to be that migrants on the way to south of the Sahara are using central sites to get ready, while many of those around the coast are checking out sites for suitability to stay all winter.

MOROCCO

The literature on Morocco does not describe large numbers of common sandpipers being seen. Dick & Pienkowski 1979 give weights for 34 caught during a general study of waders in Morocco: 23 of these were juveniles, but the mean weight of all birds was only 43.1 grams. The differences between August and September and juveniles and adults were insignificant, and all the birds were light. The general observation on most of the species that landed was that they appeared from a great height and looked hungry and tired.

In a survey of 40 km of the Sidi Moussa-Walidia coastal wetlands (including tidal lagoons, saltpans and brackish marsh) where a monthly count was done from March 1994 to Feb 1996, thus covering two southward passage seasons, only two common sandpipers were ever recorded; they were in winter (7 November 1994 and 12 January 1995) rather than passage season, and among the roughly 10,000 waders present (El Hamoumi & Dakki 2010). They did note that back in 1971 there had been 40 seen between August and September, which was clearly remarkable. When I visited Morocco in October 2001, I watched a few at Oued Massa where two were disputing over a bank, looking as if they were establishing winter territory, and a similar disposition of five at Oued Souss, looking more territorial than migratory. As mentioned at the start of this chapter, on 16 June we had a ringing recovery from Nador, on the north coast of Morocco, of a breeding bird that had been in England a few weeks before, and this is consistent with a scattering of recoveries from most European countries. But considering there are over a million individuals en route from Europe to West Africa, it appears that Morocco is not a regular feeding link in the chain.

ARE ALL STOPOVER SITES OF THE SAME QUALITY?

The review that we have just been through clearly implies that some places are stayed at for long enough to put on weight (e.g. salt-marsh and sewage works) while at others the birds just rest and feed a little and move on. Many people can see them on open reservoir edges, but are these places very good for migrant sandpipers? Perhaps a clue can be got from a look at the results published for Eye Brook reservoir (Mason1984) whose common sandpiper profile through from July to September is summarised in Figure 21 above. Data for 15 species of wader were collected from July to October over 23 years from 1957. This reservoir is central in England, and large numbers of British sandpipers must fly over it, but during July hardly any stop off there; when the numbers peak at Eye Brook in August, they are either extremely late Scottish juveniles or birds from Scandinavia. The total number of common sandpipers recorded there was 2,128, while there are also 1,986 records of greenshank and 2,748 of ruff (*Philomachus pugnax*) which is a good clue that most waders there are from Scandinavia. The 2,128 recorded over 23 years is roughly a 100 a year. These are spread over 90 days, and even at peak times this is only a few (say two or three individuals) being present on a typical day in August. The perimeter of the reservoir is given as 8.5 km. In mid-July one might find that number of common sandpipers feeding up in 0.1 km of salt-marsh channel. There are similar low numbers and late occurrence around the reservoir that I watch in Sussex. I thus argue that reservoir edges, though often easy to observe, are poor-quality habitat for passage birds, and used for resting rather than serious feeding.

So, there is a consistent picture of some adults looking for good places where they can feed for around two weeks. The times deduced from re-sightings in good German and Belgian sites in the 1960s and 1970s were 12–18 days, and the geolocator bird from Scotland stopped off in south-west Scotland and England for 14 days; similarly those at the Lune mouth. The weight gains found have been typically 1.5 grams per day, so getting up to 80 grams on a good site is entirely reasonable; at almost every site some birds near that weight were recorded. Those birds can then get south of the Sahara. All British common sandpipers and a proportion of others use this strategy: the route from the British Isles is mainly over the sea where there are no predators, and so the loss of manoeuvrability generated by high weight may be unimportant.

Across Europe there are many places that they can stop at and feed up enough to go another few hundred kilometres. Many of these sites may be used by only around 100 individuals. For those going overland across Europe the balance of risks will be more in favour of less weight and a shorter move to the next site. The studies show that when they reach Spain they can easily put on enough fat to fly across the Sahara.

Thus common sandpipers use long hops, short hops and everything in between.

ACROSS AFRICA TO NON-BREEDING HOME AREAS

The general synoptic weather chart from Iberia to Senegal is favourable (Elkins 1988) and selecting the right time to start should enable a safe crossing. There are pools and around 80,000 square miles of oases if they need a stop, and they are reported to do so.

On the eastern side they may follow the Nile, but I have found no records of large numbers being seen to do so. One near-death experience was reported of a bird at 30 grams sat in the middle of the road in Ethiopia on 8 August. After being kept safe overnight and then placed in a good feeding place next day, it recovered and over ten days increased to 43 grams (Hillman et al 1986). On his photographic expedition up the River Luwegu in the Selous game reserve in Tanzania, Matthiessen 1981 noted a common sandpiper during early August 1979 with the comment that 'it was a bird that had not bothered to return to its breeding area'. The view was widespread that waders did not breed in their first year and stayed in Africa; the thought that they could get back from breeding so soon as August was unbelievable. Of course we now know that British breeders are back south of the Sahara by the end of July, and there is no reason why Turkish or Russian breeders cannot be back in southern Africa in August. Indeed in his review of their time in Zimbabwe, Tree 2008 points out that they are the first Palearctic wader to arrive, and the main arrival may (in the right conditions of water level) start in early August. The peak arrival is during September, and birds are on the move all the time as water levels change with the seasons in different inland parts of the continent. The recoveries of birds that have been ringed in the eastern half of Africa are in Russia (Viksne & Michelson 1985).

ASIA AND AUSTRALIA

The passage southwards across Asia appears to be essentially the same, though the barrier of the Himalayas adds to that of deserts. In Hotker et al 1998 there are several papers on wader movements through western Asia, and the common sandpiper appears to move similarly to through Europe; down the Caspian Sea, also through Kazakhstan with 116 per sq. km on a sewage farm in August and some retraps in later years of ringed birds. However, in the far east, around where the Amur River joins the Sea of Okhotsk, they were quite scarce as there are few breeding north of there, but the phenology reported on the mudflats of the Penzhina river estuary in Kamchatka showed a steady flow through from 25 July to 23 August in 2002 and 2003 (Gerasimov in Straw 2005).

Anthes et al 2004 studied the waders using a small island at the north-west end of Lake Baikal (53°05'N 106°51'E). This island was 1 km long and around 10 metres wide, consisting of shingle that was swept there from an incoming river. The dates of occurrence are much the same as that in northern Britain (see profiles in Figure 21). This suggests that the fairly local breeders are on the move. A brief mention of them on the Selenga delta (52°20'N 106°30'E) at the south side of Lake Baikal says records are spread from mid-July to mid-September without any clear peak passage (Fefelov & Tupitsyn 2004). This large delta thus appears be an attractive place for migrants from a larger area to prepare for migration onwards. The delta has hundreds of thousands of migrating ducks, and 46 species of wader totalling tens of thousands.

The movements of those in India are not demonstrated by ringing recoveries. The arrival in Sri Lanka appears to have happened by late August (Serasinge 1992). It would appear likely that many of those in India breed in the uplands to the south of the Himalayas eand move further south for the winter, but they may be joined by Russian birds.

Phil Round sent me his ringing records from sites in Thailand (described in his paper on stints in Round et al 2012) where the wintering population was preceded by migration through August and September. The weights of the passage birds are lower than those in mainland Europe (see Table 2.2 above) suggesting they have an easy passage in small stages down the food-rich coasts of south-west Asia. Cramp and Simmons 1977 describe them as common through Malaysia. Lane 1987 describes them as common on migration through Japan, China and islands north-west of Australia. He also gives a ringing recovery from the Philippines back to the Amur River in Siberia. There is also a ringing exchange between far eastern Russia and Borneo.

The number that reach Australia is estimated as around 3,000 (Geering et al 2007), and they may be seen mostly on the coast, but migrant sandpipers are recorded regularly inland. Thus during an atlas project from 1977 to 1981 there were 174 inland records (Lane 1987, Table 4.2). There are a few records in New Zealand, but it is fair to say that those going further south than northern Australia are an insignificant part of the global population.

SPOTTED SANDPIPERS ACROSS THE AMERICAS

The spotted sandpiper southwards migration has the same features as those we have seen for the common. The female stays somewhat longer on the breeding area because it is more adapted to multiple clutches, but still leaves as soon as the last chicks are part-grown and thus before the males. The birds migrate in small groups or singly. Captive birds display restlessness in late summer during high pressure fronts and north-west winds, and also have large weight gains of up to 100 per cent (Oring et al 1997). They are observed on passage in all areas and on all sorts of water bodies, like those used by common sandpipers. A detailed profile at Long Point from Bradstreet et al 1977 is shown in Figure 21, and general times quoted for various states in BNA are as follows.

Table 2.4 Migration through northern America

	Early birds	Peak birds	Last birds
British Columbia	mid-July flocking	mid-August	October
Alberta		August	mid-October
Saskatchewan			10 Sept
Manitoba			18 Oct
Quebec			12 Oct
Pacific NW	late July	mid-August	early Oct
Minnesota		25 June–23 July	
Ohio	12–18 July	25 July–25 Aug	early Nov
Central Texas	27 June		22 Nov
Florida	late June		mid-October

Figure 22: Greg Lavalty sent this superb picture showing the key features of a spotted sandpiper skimming the water in Texas.

Clearly the fall migration is protracted, and as with common sandpipers the main passage is in August; but stragglers/explorers may turn up much later and as we will see in Chapter 3 they may stay in coastal areas and warmer southern states for the winter. As in European countries (e.g. Spain) there are states (e.g. Texas, Figure 22) which have both breeding and wintering birds as well as passage. But again the detailed habitat used is often different, so that separating passage birds is generally practicable.

There are plenty of records of vagrant spotted sandpipers in north-west Europe, and these start in mid-August and continue into November (and many stay to the next year); the arrival timings are not dissimilar to those of common sandpipers arriving in Africa. This is reasonable, because the distances are similar. From 1950 to 2011 the British Birds Rarity Committee listed 179 occurrences.

The Canadian Atlas of Bird Banding reported that one bird which had been ringed in August in New Jersey was recovered three years later in the Mackenzie delta in north-west Canada (68°50'N 136°20'W). It thus seems plausible that some birds from the far north use a similar strategy to that of white-rumped sandpipers (Harrington in Able 1999): they fly to the east coast and then cross the sea to the northern coasts of South America. Some then get blown off course and so turn up in Bermuda, or in Europe – or in the most extreme case Tristan da Cunha, which had one collected on 5 February 1952 whose skin is now in the British Museum of Natural History (BMNH) at Tring.

The majority of birds are probably less adventurous as they work their passage south. Alerstam 1990 shows the sea-crossing distances from various places on the east coast of North America across to the coast of South America decreasing from 4,300 km from Nova Scotia to Suriname, 3,200 km from Virginia to Venezuela, 1,900 km Florida to Colombia (since this crosses Cuba there are resting opportunities halfway) and 1,100 km Louisiana to Mexico. Individuals migrating further west can avoid crossing any ocean at all as they can work their way down through Central America.

The first reach Suriname in July (Spaans 1978). Oring reported that one of his Minnesota breeding study birds was observed in Guyana in September. One ringed in August in New York was in Martinique 25 days later, while one from Illinois was in North Carolina in September (BWP). Together with four birds from Ontario, one of which was found nearly due south, one due east and two south-east, this pattern shows a fanning out of birds as they migrate, reminiscent of common sandpipers.

On flooded grassland in southern Colombia at around 1,000 metres above sea level, spotted sandpipers were present around the pools from August onwards into November, and were the second most numerous shorebird after upland sandpipers (*Bartramia longicauda*) which used the grassland. The reduction of spotted sandpipers in November was put down to increased disturbance from agricultural and recreational activities (Ayerbe-Quinonez & Johnston-Gonzales 2010).

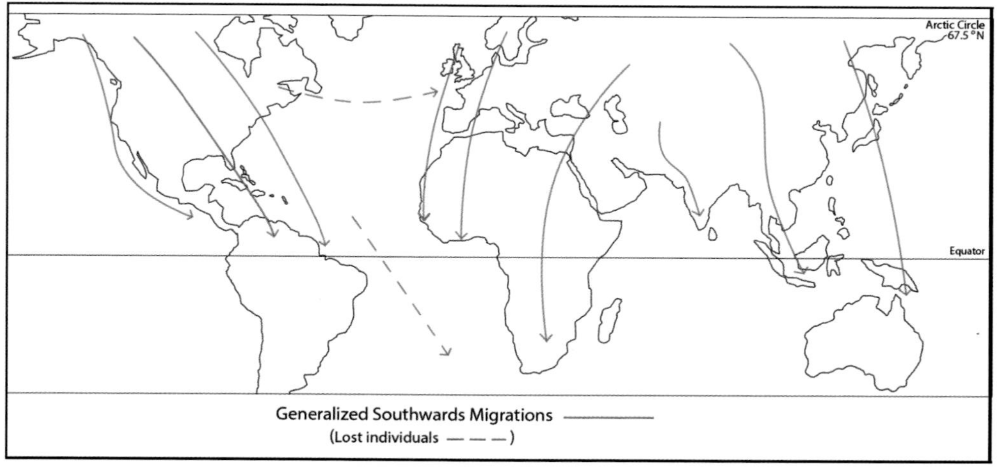

Figure 23: A generalised picture of the southwards migration pattern based on all long-distance recoveries of both species published in various books.

3 BEHAVIOUR IN THEIR WINTER HOME

The majority of individuals spend the majority of their life in the tropics where they live from August to April. Indeed the first common sandpiper to have a geolocator fitted got to West Africa on 29 July; most African countries report first returns in July, and similarly with spotted sandpipers in South America. The geolocator bird did not leave West Africa until 10 April, so was there for over eight months (Bates et al 2012).

In this chapter we look at where they are during the northern winter and what they do. Their needs are to survive (find food, avoid predators, avoid diseases) and to moult into a good state for migrating north. There are some parts of Europe that have common sandpipers all year, and some parts of North America that have spotted sandpipers all year. But usually the habitat used by the birds in winter is different from the breeding habitat.

My interest in winter behaviour was greatly enhanced by watching a spotted sandpiper on a beach near San Diego, California. It was a big beach with the tide out. On the damp sandy mud were a few shorebirds, one being a spotted sandpiper. Then another one appeared and they started to squabble. One called, raised its wings and got so agitated that it twirled around through 360 degrees, something I have never seen a common sandpiper do. This was an enormous beach, and a mixture of other shorebirds

 Figure 24: Two spotted sandpipers squabbling while a plover looks on.

just got on with feeding. This territorial antagonism seems widespread while feeding in winter: in several countries and habitats while I have been watching one sandpiper feeding (often with other species) if another of its own species comes into its space there is a dispute. However as darkness comes they are often seen to gather into small flocks for roosting.

MOULT

Other than surviving, their most important activity in the wintering areas is to moult and be in tip-top condition for the big effort of migrating north, breeding, then migrating back home again. The juvenile small feathers that cover the wing are very distinctive in having golden patterns (Figure 15), and patterns on the first-years' tertials differ, enabling them to be identified (Meissner et al 2015).

The most comprehensive data on moult are those from Africa south of the equator (where it is summer, so a more appropriate adjective is 'non-breeding' rather than wintering). Tree 2008 summarises results from the primary moult of 271 adults and 193 first-years in Zimbabwe caught between 1972 and 1993. For adults the moult is the same as in most birds; the innermost primary is shed first, and the moult progresses steadily outwards with typically one feather just lost and another nearly full grown. This starts in August or September and may take up to five months. There were some exceptions in birds that were found nearly complete in November, and it was postulated that these were birds now in their second year that had been unable to migrate all the way back to Russia and had stayed in Africa as non-breeders, so taken the opportunity to start moult early. This is a reasonable deduction from the variability of first-year moult. First-year birds may start their moult at any feather from primary 1 (innermost like the adult) to primary 6 and thus finish with a reduced set of new outer feathers (in one case a bird started at primary 9, so only had two new primary flight feathers). In a few cases moult started in the middle and went both outwards and inwards, which was called centrifugal.

Table 3.1 The percentage of first-year birds in Zimbabwe that started moult at primaries one to nine

Starting primary	1	2	3	4	5	6	7	8	9
Percentage	2.6	8.3	19.7	49.2	17.1	2.1	-	-	0.5

These first-year birds' moult in Zimbabwe generally started in November or December, but there were extremes from October to January. The earliest completion of a partial moult was on 28 February, but all must finish soon thereafter so that they are able to start their long northward migration to breed.

Pearson 1974 reporting on birds caught in Kenya, mainly at sewage works around Nairobi, had similar results, with adults usually starting in October and finishing in February while first winter birds started in January normally from the fourth or fifth

Figure 25: The problem of not renewing feathers is that they wear out. If this individual had not replaced the outer six primaries it would hardly be able to fly; photograph of a common sandpiper in August sent by W. Meissner from southwards migration studies in Poland.

primary and finished in March. He also commented on the moult of the secondary feathers: they are renewed rapidly, somewhat randomly and sometimes not at all. Body moult is going on all the time, with smarter bronzy feathers for breeding and duller brown ones for the rest of the year. Wing coverts gradually change from the juvenile ones of Figure 15 by wear and by replacement with adult feathers, and are generally of little use for ageing beyond February.

The wear that is sustained by four juvenile primary feathers that were not replaced is shown in Figure 25, taken in Poland during the southwards migration in August, with inner primaries that are over a year old, on birds that are on their third long migration.

Very detailed studies of the moult of wood sandpipers (*Tringa glareola*) in South Africa (Remisiewicz et al 2010) show again that the first-year birds replace a few primaries, the number depending on the starting date of their moult. Thus 63 per cent of them, starting early January and taking 73 days, replaced four feathers, while 29 per cent started earlier, in mid-December, and replaced five or six. They all finished in the fourth week of March. The flexibility of primary moult in a range of other Palearctic waders is comprehensively covered in Remisiewicz 2011. An extreme example of the flexibility of waders is that dunlins in the east of Russia start moulting while incubating and take about 50 days on the breeding ground, while those from the west of Russia and northern Europe migrate to western Europe, and then spend about 100 days moulting before moving on to their final wintering haunts.

The following data have been provided by people who have caught a few common sandpipers as part of studying other birds, and largely make sense within the framework of flexibility seen in the other studies.

Round et al 2012 studied stints in Thailand among mangroves, and also caught sandpipers. Some stayed all winter but many passed through. In September seven of nine adults were moulting while none of the 16 first-years were. By November ten adults had completed moult and one had arrested moult, but four first-years were still not moulting. Only three were caught between December and March inclusive, of which two were in moult. In Gambia some sandpipers were caught in March 2011, December 2011 to January 2012, February 2013 and January 2014 (Blackburn, priv comm.). Of 30 birds, 23 per cent were adults and none were recorded with active moult, which is consistent with them having finished by the end of the calendar year if they started in August. The first-year birds showed a variety of stages of moult. Some were consistent with a steady moult from primary 1 starting late in the year and completed by March in time to fly north. But one in March 2011 had just two new primaries (6 and 7) and two birds having started at primary 1 appeared to have stopped before completion. Pienkowski et al 1976 found that three juveniles out of twenty-three caught in mid-September in Morocco were already halfway through primary moult. A first-year bird that I caught near the south coast of England in November had already renewed primaries 1 to 7 (Figure 26). It was not on passage, as it stayed there through the whole winter; birds caught on passage in Europe are rarely in moult. Brown 1973 reported one that was caught at Wisbech sewage farm on 24 September 1964 with new primary 1 and half-grown primary 2, and the same bird was retrapped on 23 August 1969 with three new inner primaries, primaries 4 and 5 growing and the outer five primaries old; he also noted

Figure 26: Unlike some birds, they only have a couple of feathers growing at a time; this first-year bird in Britain was already on its eighth primary feather by 10 November.

five others (out of 200) with some moult. In Scotland in spring, birds have also been caught which had only replaced a few outer primaries (Nicoll & Kemp 1983 and Meissner et al 2015), but where they had spent the winter is not known.

Spaans 1979 caught 658 spotted sandpipers in Suriname between 1970 and 1977; he found that adults started moult from August to December and estimated they took four to six months. First-year birds started between November and February, and some retained old inner primaries (up to five) like common sandpipers.

For many birds their food is abundant and lasts long enough to allow a bird to breed and to moult before their migration south, but for sandpipers food is already getting scarce on the breeding territory in July, so the female gets away even before the young fledge, and all sandpipers get away as soon as possible. Important decisions are needed on moulting because the bird needs a good food supply to grow the new feathers and while it has less than its full set of feathers its flight is impaired. But it must replace them eventually if it is to breed. I suspect that part of the importance of wing-raising displays on the breeding area is to show potential mates and competitors that the wing is top-notch. Thus a first-year bird with a full set of new primaries shows its potential mates that it is better than a bird that only has a partial set.

Thus the adult gets back first to a known winter home, and gets on with moulting at a speed that is consistent with the food supply so it can focus on survival and preparation for the next summer. For a first-year bird it is more difficult. The best feeding areas have been taken by adults, and its first job is to survive and find a home. If it has ended up in southern Africa it needs to decide whether it has enough food and time to replace all its feathers, and if not all, then where to start. Primary 4 seems a favourite choice and will give it enough new feathers. The same decision will have to be made in other parts of the range. If it gets a good territory (maybe in mangroves) it can moult like the adults, and starting just a few weeks later will still complete well in time, as it will also have a later schedule for leaving than those that need to do the much longer journey from southern Africa. When the good territories have been taken then the weaker birds are faced with the same decision as the southern birds.

The few birds that winter in coastal Europe may make the decision that whereas those that are going to Africa use the plentiful food supply to amass energy for migration, their best option is to get on with moult while the food is good and the weather is mild. This gives them perfect plumage to insulate them in the cold, wet and wind (the same reason that other shorebirds moult promptly in northern estuaries). It could be the other way round; that some juveniles start to moult due to a defect and then find that they are unable to migrate far, and that causes them to stay in the north.

WHERE THEY ARE

Standard books give distribution maps that show common sandpipers uniformly over a vast area. I try here to give a more detailed distribution of their settled winter distribution as well as looking at what they are doing. Similarly in the Americas the distribution of spotted sandpipers is shown in guides as uniformly distributed over a vast area, though the numbers in the far northern and far southern parts of the winter range are a minuscule fraction of the total.

DENSITIES IN MANGROVES AND ELSEWHERE

As we go around the world in the next paragraphs we will see that places where high densities are found are in mangroves. When large flocks are reported they are close to mangroves. The estimated world population of common sandpipers is about 3 million and the area of mangroves in the old world is about 105,000 sq. km (Spalding et al 2010). Thus if the density was 30 per sq. km then all could winter in mangroves. The latest estimate of spotted sandpipers is 660,000 (Andres et al 2012) and the area of mangroves in the new world is 45,000 sq. km so at a density of 15 per sq. km they could all fit in mangroves. Some of the estimates of numbers per square kilometre in mangroves as we go around the world are into double figures. Clearly there are many other habitats with wintering sandpipers, and part of the purpose of this following world tour is to judge how important they are. There are several different species of mangroves, and it may be that some are more useful to sandpipers than others, either directly or because of the other species locally using them. The non-mangrove general areas across Africa, India and the East, where common sandpipers are described by words like widespread, abundant and common, covers around 3 million sq. km, so even at one bird per square kilometre there is room for them.

A TOUR AROUND THE WORLD

The spotted sandpiper is included in the field guides for every country in the Americas, indicative of their ability to survive almost anywhere. It is a common experience that any river bank or large pool with open edge may have a sandpiper. But turning these observations into numbers is difficult, given the transitory state of many habitats. Whilst on a tour of Costa Rica I kept detailed notes of where I saw them and questioned guides on places to see them. They were abundant in the mangroves of both coasts but also along the rivers through the interior: my guess for the country population was 10–20,000. All the countries fringing the Caribbean and the islands have good numbers. Audubon midwinter counts show them well distributed across the southern states of USA and well up the west coast. BNA reports them as regular on Vancouver Island. All the countries in South America have some, but they get less and less going further south. In north-eastern Brazil survey they were one of the most frequently encountered but not in large numbers, and they were in the inner estuary rather than out in the open (Cardoso et al 2013). Birders' lists from the enormous Pantanal wetland, however, rarely have them: I suggest that most birds that decide to push on beyond the mangroves fringing the north coast will find somewhere on the tributaries of the Amazon, and a good spattering of records across Brazil can be seen on WikiAves. Spotteds going further are essentially vagrants, and they have even been reported in Tristan da Cunha (as mentioned in Chapter 2).

In a survey of Suriname's shorebirds (Spaans 1978) the main areas for spotted sandpipers were the mangroves. He did regular counts along a 6 km stretch of mangroves and showed around 30 birds using that stretch from August to February (5 per km). He also did a regular count at the mouth of the Krofajapsi Creek draining a mangrove area, from which large numbers emerged in the evening to roost away from the mangroves. These counts were very variable, but essentially 200 birds were about from late July through to April, with a peak of 460 in January. In contrast, 8 km of sandy beach only had numbers in August and September, and the open muddy estuarine area had none. The total population of Suriname was estimated as 10–50,000 in a country that has 510 sq. km of mangroves (so potentially 20–100 individuals per square kilometre). In mangroves on the Pacific coast of Colombia a flock of 150 was noted (Johnston-Gonzales et al 2006).

Moving across to West Africa, a comprehensive study is reported for Guinea-Bissau (Zwarts 1985 and 1988). This looked at which birds were feeding on fiddler crabs, *Uca tangeri*, and the differences in wader populations on different substrate types, then went on to use this to estimate total wader numbers in intertidal areas for Guinea-Bissau. The results extracted as relevant to our study are:

Table 3.2 This shows that the most popular area was the mixture of stone and mud, and this survey area was along the Rio Grande de Buba and Canal do Porto that are flanged with extensive mangroves. To extrapolate numbers to the whole country, the data were interpreted as a density that depended on sediment score, a subjective number ranging from 1 for coarse sand, to 3 fine sand, 5 muddy sand and 7 soft mud, with clay content logarithmically increasing from about 1 per cent at score 3 to 20 per cent at score 7. An estimate of 9,100 birds on the intertidal flats was derived. To these the author points out must be added those actually in the mangroves 'some tens of thousands of waders, mostly whimbrel (*Numenius phaeopus*) and common sandpiper'.

	Mixture of stone and mud	Soft mud	Mud and sand	Sand
Area surveyed: ha	945	875	5615	875
Waders per km²	131	746	951	333
Common sandpipers per km²	28	18	1	0
Per cent by number that are CS	21	2.4	0.1	0
Per cent of total bird mass as CS	4.7	0.1	0	0

Moving to Guinea, the winter waders were counted by Trolliet and Fouquet (2004). In this study many of the counts were made from a boat and the counts were done from 1999 to 2002 over four winter visits (December or January), each taking two to three weeks. In total they counted 140,992 waders of which 1,782 were common sandpipers. They used their counts to extrapolate to a country coastal population of 8,330 on coastal mudflats and 15,000 in mangroves. They counted at mid-tides because at low tide they could not see the birds over the large areas, and at high tide many, especially sandpipers, were roosting inconspicuously in the mangroves. In the case of coastal mudflats they sampled 28 per cent of the area, so the amount of extrapolation was considerably less than for mangroves, for which the sampled length of channels was about 2 per cent. The importance of mangroves to sandpipers is indicated by the fact that on the mudflats 1.7 per cent of the waders were sandpipers, while in the mangroves 37 per cent were. The area of mangroves in Guinea is 2,033 sq. km (Spalding et al 2010), so at face value the 15,000 birds in mangroves are at a density of 7.4 per sq. km. A value of 9.45 per sq. km had been reported earlier from an area of mud in the mangroves (Altenberg & van der Kamp 1988).

Sierra Leone wintering waders were assessed by Van der Winden et al 2009, with high coverage of mudflats and creeks in mangroves. Again the percentage of waders that were common sandpiper on coastal mudflats was low at 3 per cent, while in mangroves it was 23 per cent. They gave results separately for creeks less than 100 metres wide (2.09 common sandpipers per km) and those that were 100–300 metres wide (1.27 per km). The increasing density in smaller creeks leads to asking whether up creeks too narrow for boats there are lots of unrecorded birds. The total population was assessed by extrapolation to be 4–5,000: this was lower than an earlier estimate by Tye & Tye 1987, who gave 4,700 just for the Sierra Leone river mouth and another 4,500 for Yowri Bay, but also gave a detailed count at one place (Aberdeen Creek off the Sierra Leone river). With this variation through the winter at one place, the difference between national estimates of 5,000 and 10,000 for the country is not surprising. The low numbers before November may imply that they moult somewhere else before arrival; a rise from 50 in November to 200 in March may indicate that the drying out of inland pools steadily pushes more birds towards the coast.

Moving along from these three countries which have good populations, we see that in Ghana (Ntiamoa-Baidu & Grieve 1987) the habitat is less attractive, and a population along 500 km of coast is estimated at 806. Ghana had very small area of mangroves (136 sq. km) and it is not obvious whether 'coast' included mangroves anyway; inland in Ghana there are large reservoirs. As in neighbouring countries sandpipers are widespread but not exceptionally abundant in any one place, but will be seen at any river or pool (eg in Ivory Coast, Francis et al 1994) and this applies through Nigeria, Cameroon and Gabon to Congo.

In the north we reach the Sahel. *Living on the Edge* (Zwarts et al 2009) indicates that the Sahel is not a very important area, especially by contrast to those species for which the area is very important.

Table 3.3 Some counts extracted from *Living on the Edge*

	Senegal Delta floodplain (LotE – Fig 79) Bird per ha	Inner Niger Central Lake – Mali (LotE – Table 5) Max of 62 counts	Lake Chad – Chad (LotE – Table 15) Feb 2001 count
Garganey	Not on Fig 79	273,000	23,869
Ruff	0.5	90,470	146,343
Little stint	6.1	38,362	2,212
Wood sandpiper	1.3	755	2,926
Common sandpiper	0.2	74*	None on list

* This count was in August, so even this small number were probably on passage.

The Senegal delta at the western end of the Sahel is getting close to the main mangrove areas where we started our review of West Africa, and another table (17) in *Living on the Edge* looks at the use of rice fields between November and February. In rice fields categorised as coastal and wet, the density of sandpipers was 37.1 per sq. km (compared with ruff 22.6 and little stint 10), and the authors suggest that this is associated with the proximity of tidal flats and mangroves. Overall they conclude that the rice fields only partially compensate for lost habitat in the floodplains. Rice fields well inland in the inner Niger Delta had no wintering sandpipers (*LotE* Table 19). Wherever they are, the waders rely on the rice fields having enough water. Overall the available habitat for sandpipers is an ever-changing result of weather patterns, major reservoir construction and detailed local agricultural management. On a daily basis they may move at high tides from mangroves to nearby rice fields. Further north than Senegal, one of the great wader sites of the world, the Banc d'Arguin in Mauretania holding around a million waders, has no common sandpipers. But the detail in the report of the Dutch expedition there (Altenburg et al 1982) describes how on arrival at the Port of Nouadhibou on 11 January 1980 there were three foraging on the waterline and the next day there were six resting on boats; and on return in the first five days of March there were five to ten each day. This nicely shows that they find all sorts of small niches in quite unusual places, but not among flocks of other waders on open ground. Similarly two were found in Dakhla Bay in the western Sahara (Rufino et al 1998).

Where the Sahel meets the Nile (1,000 km east of Lake Chad) is another major wetland; the Sudd, which is not included in Table 3.3 above. This is largely marsh and swamp, and with wars is now little surveyed. There must be pools and shoreline that can be used by sandpipers, but the fact that birds push on through the Rift Valley and on to southern Africa suggests that it may not be very densely populated with sandpipers in winter. Some may mix with those through the Congo basin down to the west coast. In the Congo sandpipers occur 'practically everywhere there is some sort of water' (Curry-Lindahl 1960).

In Africa, numbers down the Nile and on through the Rift Valley and on to southern Africa extracted from Summers et al 1987 are:

Table 3.4 These are only a small fraction of the estimated breeding population needing accommodation. But those countries have small areas of mangroves, though attraction to mangrove roots on passage was reported (Milligan 1979). Tanzania, Mozambique and Madagascar have around 7,000 sq. km, which could hold around 70,000 sandpipers. Few of the large lakes or the rivers feeding them, or indeed the Zambezi, appear to be included. With many million square kilometres potentially available there only need to be a few birds in every 100 sq. km to account for the western Russia numbers.

Southern Africa coasts and coastal wetlands	2000 (850 in Natal)
Kenya inland	2500
Kenya coastal	300
Sudan inland (the Nile)	1000–2000
Egypt	100s

Pearson studied them near Nairobi, where they used rivers and sewage works and pools but did not use alkaline lakes. A survey every five days from October to December 1987 around Lake Oloidien, a satellite of Lake Navaisha which had become separate in 1979 due to lowering water levels, found 80–100 every time (Lyngs 1996). But by 2006 the salinity had increased such that 250,000 lesser flamingos (*Phoenicopterus minor*) had taken over. Tree 2008 describes them being most common on the eastern side of southern Africa, which is moist, and being scarce in the west, much of which is desert. He reports that reservoirs and sewage works have attracted them, and indeed much of the ringing that has made their birds the best studied non-breeding population has been done at sewage works, where they come to roost on non-operational filter beds.

All the ringing recoveries from southern Africa have shown they come from Russia, and the direct route takes them over areas like the Caspian Sea and Arabia. In Hotker et al 1998 a few are noted wintering round the Caspian Sea, and in a survey in Iran on the Ramsar site of Gomishan on the Caspian they were recorded every month from September to February, peaking at 13 in January (Ghaemi 2006). The gulf coasts of Iran have estuaries with mangroves that are Ramsar sites and IBAs, so should have a few. But midwinter counts in Saudi Arabia of 27,783 waders had zero common sandpipers (Zwarts et al 1991). Simmons 1951 reported on their winter territorial behaviour in north-east Egypt, using a coil of wire as a territory boundary.

Moving west to the Indian sub-continent, Serasinghe 1992 counted birds that came in ones and twos at dusk to roost on an abandoned barge in a fairly urbanised area of Colombo in Sri Lanka. Observations were made from end of August to early April, when 11–17 birds were counted. Ward 1999 reported 70 in winter for part of the Indus delta in Pakistan. On a holiday in Goa in December 1997 I found that many ponds and wet paddy fields had one, and I also watched a few feeding on a sandy shore. In January 2006, down central India from Bharatpur, they again were dotted about, e.g. on the river by the Taj Mahal. Birders visiting the Sunderbans report them as common. Standard texts on the sub-continent say that they are widespread.

The greatest concentration of mangroves on the planet is in the Far East, and personal observations, internet searches and other communications suggest that mangroves are well frequented by sandpipers. Thus on the birdingneverstops blog was a record of 200 coming out of mangroves to roost on boats. While surveying in Myanmar for spoon-billed sandpipers (*Eorynorrhynchus pygmaeus*) in January 2015, Nigel Clark (priv comm.) found flocks of around 100. While on a boat in mangroves near Krabi in Thailand I counted ten feeding along around 2 km of channel.

Those lurking down the tropics from southern China to Australia are not well enumerated, but if they were at 10 per sq. km of mangrove, there would be space for around 600,000. Lane 1987 gives a summary of where they are found. It is also reported that considerable numbers of non-breeding birds stay during the breeding months. I do not know whether this is a real difference from elsewhere or just the disbelief that birds are absent for so short a time to breed. Contrary to the approach which I have taken in the last two paragraphs (implying several hundreds of thousands), an expert appraisal is of only around 30,000 in the East Asian–Australasian flyway (Bamford & Watkins, in Straw 2005).

For Australia, Geering et al 2007 give a detailed map and an estimate of 3,000 common sandpipers. The distribution is described as 'most abundant in mangrove inlets but can be found on narrow muddy margins or rocky shores of a wide variety of coastal and inland wetlands'. Their map gives the location of every sighting from two databases covering 25 years, and there are records from all the way around the coast and a few inland. However in the text (and also Lane 1987) the north-east is described as their main area; this is consistent with that area having the main concentration of mangroves and it being the shortest sea crossing as birds move south. In the islands immediately north of Australia, sometimes called Wallacea, which are not ideal for waders, a brief review showed common sandpipers were the most commonly recorded wader (White 1975). I have assumed the scattered records in the centre of the Australian continent are essentially vagrants, and the occasional vagrant also gets to New Zealand (less than eight birds per decade said nzbirdsonline).

In the BMNH, there are plenty of skins of common sandpipers from the Solomon Islands, and on Vanuatu they are described as the most abundant wader (Bregullqa 1992). BNA reports occasional records of spotted sandpiper on the Hawaiian islands, but neither species is abundant on very remote Pacific islands. All over the world there are records of them on rocky shores in very small numbers, but the way they utilise this habitat and whether they hold winter territories in such surroundings is unclear. While searching the internet I came across a birding blogger (John Seymour) who said he had seen common sandpipers in more countries than any other species, so it has a good claim to its English name – though 'widespread shorepiper' may have been a good name to give it, and *flussuferläufer* ('riverbank runner') even better.

An attempt to summarise the winter concentration onto a map is shown as Figure 27.

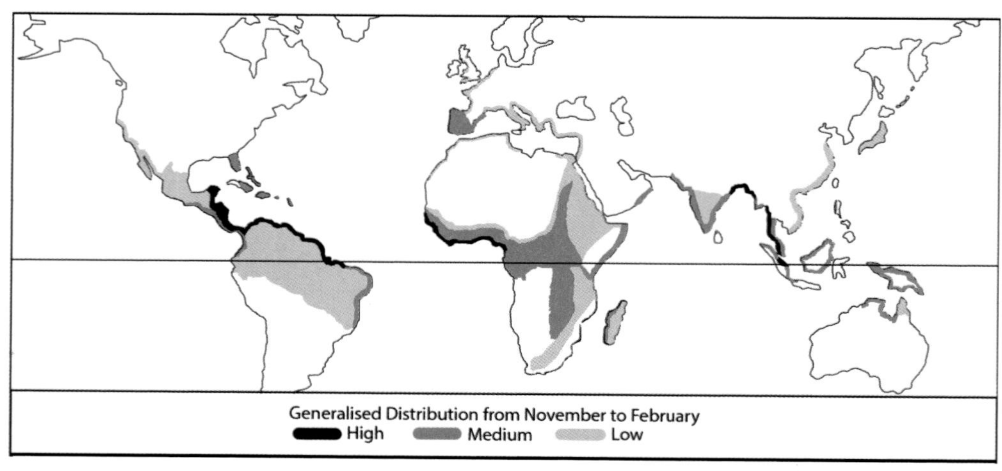

Figure 27: This is my attempt to show non-breeding areas in three categories. The black is regions of potentially high density (10 per sq. km) in mangroves, the red is a middling density where there are rivers and pools (a few per sq. km), and the pink is where regular but sparser records are. In blank areas they may be found, but are not numerically significant. The densities multiplied by areas are consistent with a population of around 3 million commons and 1 million spotteds, but this clearly cannot be used as a census estimate.

COMMON SANDPIPERS WINTERING IN BRITISH ISLES AND THE REST OF WESTERN EUROPE AND DOWN THE ATLANTIC COAST TO MAURETANIA AND THE SOUTH SHORES OF THE MEDITERRANEAN

In 1976 in the British Isles it was reckoned that about 100 wintered, and in 2013 about 73 (Lack 1976, Balmer et al 2013). In a review of waders around the Mediterranean, Smit 1986 gave estimates of 10–20 for each country, except Italy which had 700, although Perco 1984 had described 1,500 just in the northern Adriatic part of Italy. These are classic manifestations of their low profile in most countries; they tend to avoid areas popular with other waders, and they are difficult to enumerate and sometimes ignored. A related problem is that one in November can be dismissed as a late southbound migrant and one in February as an early northbound migrant.

Arcas did studies in north-west Spain on the Miño estuary bordering Portugal (where it is called the Minho), and on a fairly industrialised estuary near Vigo. He studied their food (covered in Chapter 5) and found they were territorial, typically using less than 100 metres of a channel where they fed for a high proportion of the day. Their territorial defence took up a significant part of the day while they established their territory in November, but once established all the birds respected their neighbours' space much as they do on breeding territories. On non-estuarine coasts of Galicia, Spain, 0.33 per km were reported (Dominguez & Lorenzo 1992), and around the coast of Portugal 324 were found (Lecoq et al 2013). Velasco 1992 did a survey of 1,343 km of rivers in the middle of Spain, and through the winter averaged 0.46 per km. If that is representative of all the rivers in Iberia, that would imply a few thousand birds inland. The Spanish population is thought to be largely resident throughout the year (e.g. Diez & Peris 2001). They are also recorded along the south side of the Mediterranean (Spiekman 1992). The situation down the coast of north-west Africa seems to be similar; a few, but difficult to count.

If 700 in Italy was the best number in the 1980s (Smit 1986) I would suggest another 700 around Greece and the non-Italian Adriatic, another few thousand around the Spanish and Portuguese coasts (plus the few thousand inland on rivers), another thousand or so up the habitat close to the Atlantic coast of France and the British Isles, and more down through Morocco to the Canary Islands. But they are so mobile and often invisible that this total of around 10,000 is admittedly a guess. However it is important to know that they are flexible enough to survive in northern Britain. While I was writing this, Brian Bates emailed to say one was photographed on the Spey in northern Scotland on 10 December 2016.

When I moved south in England, away from breeding areas, I took up watching them in winter near the south coast. 'Near' is an important word. The typical place is a few kilometres up a tidal river (and not counted by estuary bird counters), where there are few other birds and they lurk under banks; I have walked a stretch where I know there are a few, yet seen none. While I have watched one under the opposite bank, walkers with dogs have gone past very close; the bird ignored them and would have been invisible to the walkers (or to me if I had been on that bank). During a very cold spell of weather on a day with a bitter north wind but a clear sky I watched one at high tide that was on the south-facing bank, snuggled in dry grass. When it flew off to feed as the tide went down, the depression it left was warm to touch. They have no competition for food, as they are usually the only shorebirds there. Their peck-rate and visual success with occasional large items suggests that they get food reasonably easily. They start being territorial in October, and one I colour-ringed on 7 November used the same territory of about 500 metres of river until at least 10 April, when I last saw it. Each of the main rivers of Sussex – the Cuckmere, Ouse, Adur and Arun – may have a few or may have none. Occasional sightings along the Sussex coast in winter may be of one moving from one river to another.

To get an idea of where they come from I sought permission to look at the Euring database (du Feu et al 2016). There are 40 recoveries in France between mid-October and mid-March, spread around from Corsica along the south, up the Atlantic coast and la Manche (the Channel) up to the Belgian border. These birds had been ringed with the bulk of migrants caught in Scandinavia, Germany, Netherlands and Belgium, mostly in August. The dataset released to me did not include several countries, so I took in addition, from the Polish passage (Meissner 1997), two that were found in December (in South Spain and Italy) another in January (on the Mediterranean coast of France) and one from Czechoslovakia (Glutz et al 1977), plus one in Italy in November.

Three recoveries in the Euring database were ringed as chicks, so of definite provenance.

Figure 28: Tidal River Ouse near Lewes, Sussex, England; end of wintering territory for colour-ringed common sandpiper.

Table 3.5 The place of birth of wintering birds in EURING database

Norway	68°51'N 23°5'E	26 June 2002	**France, Dunkerque**	50°57'N 2°12'E	9 November 2003
Finland	66°08'N 28°45'E	24 June 1992	**France, Corsica**	42°41'N 9°18'E	4 November 1992
Sweden	58°47'N 12°31'E	4 July 1960	**Spain, Tenerife**	28°24'N 16°33'W	10 November 1960

As well as the Dunkerque bird, three others in north-east France (two of which were recovered in January) had been ringed in the first half of August, two of them in the Netherlands and the other in Sweden. It thus seems reasonable to deduce that British wintering birds are a similar random set of dropouts from normal migration. None of the birds that were first ringed in Spain in winter were recovered outside Spain, and some were caught again in the same place in later years, the longest being 12 years later. So the whole picture is that some birds, which are residents, winter around Iberia (and probably other Mediterranean countries), and then others, in both Iberia and the rest of Europe in winter, are a random selection of individuals dropping out from normal migration. The very small proportion of birds caught on passage which are also moulting may be the ones that stay, but none of the birds in the winter recovery set had had moult recorded on first capture, so this is an untested hypothesis.

One British recovery that has intrigued me is a chick ringed in Finland (60°19'N 24°38'E) on 17 June 1959 and recovered in Devon (50°37'N 3°25'W) on 9 June 1960. It is the only record from Finland in Britain; it may be that it was a conventional bird that had gone to West Africa the previous autumn but on return was lost or adventurous, but more likely it is one that ended up wintering on the Atlantic coast and decided to try breeding in Devon. We'll never know. But we do know that the species is variable and flexible, and so is the weather. Another indication of the origins of birds drifting across to British Isles late in the year is a recovery on Fair Isle on 11 October 1981 of a juvenile ringed in Norway on 2 August.

LOYALTY TO NON-BREEDING SITES

In the British Isles it has been noted that sites tend to have a bird or birds a few years running. There is one ringing recovery that would imply they are the same birds returning: one caught in January near Southampton was recovered the next winter close by. The only wintering bird I have colour-ringed stayed all winter but was not seen the next year. At four sites in Spain there were retraps in later years (near Vigo, Madrid, Malaga and Alicante). Sauvage et al 1998 reported from Senegal that 14 per cent of common sandpipers ringed were retrapped in later years. In Zimbabwe one bird was captured twice again in the four years afterwards, and of the 418 adults ringed 15 per cent were retrapped in later years. Given a normal catching efficiency and allowing for mortality these are indicative of quite high fidelity to wintering sites. This is not surprising as this is found widely for shorebirds, but I have not found any confirmation for spotted sandpipers.

In following chapters on food and predation and other threats, we will see that they do not necessarily have an easy life on the non-breeding areas. The fact that some juveniles are unable to grow a new full set of primaries suggests that youngsters struggle, and the slow speed of adult moult implies that they do not want to have more than one feather missing at a time, suggesting that flying capability is important to avoid predators. Weights are around 40 grams, much lighter than on the breeding grounds; in a reliable tropical climate they do not need to carry any extra weight.

DOES ONE SEX OR AGE WINTER FARTHER SOUTH THAN THE OTHER?

It is reported (BNA) that female spotted sandpipers winter significantly further north than the males (based on museum studies of 51 females and 43 males). However, when I checked in the BMNH bird collection for common sandpipers down Africa, I found the following results, which suggest the contrary for common sandpipers (47 females and 56 males).

Table 3.6 Numbers of each sex dependency on latitude

	Adult male	Adult female	First-year male	First-year female
Sudan, Ethiopia etc approx 15°N	12	6	5	6
Kenya, Tanzania etc approx 5°S	16	8	11	9
Zimbabwe, S Africa approx 25°S	7	12	5	6

For spotted sandpipers the females leave the breeding area first, they are bigger, and they need to get back first to occupy their territory; so they are dominant in all three aspects, and taking good non-breeding sites further north may be expected of them. For the common sandpiper, although again the adult female gets away first and is bigger, it is the male that needs to get back first to occupy a territory. The sample above suggests that males could be preferentially taking non-breeding sites further north, though the sample is small so should not be taken too seriously.

The percentage of juveniles in the common sandpiper sample is 42 per cent and not increasing further south. This is similar to the 40–50 per cent reported by Tree 2008 in Zimbabwe from ringing studies. This is useful, as it implies that a typical pair of sandpipers produces more than 1.5 offspring that reach the farthest wintering grounds. This is higher than the productivity estimated as leaving the breeding areas in British Isles (Chapter 1).

NOT RETURNING TO BREEDING AREAS FOR NORTHERN SUMMER

There appears to be a small number still in Africa and Australia during June. The possible reasons for this are

- they are too old or sick and soon die, and would die on migration anyway
- they have some lesser illness that leads them to decide it is better to avoid the migration for one year in order to increase their long-term life chance
- they were late chicks that got a poor territory when they eventually got south, and similarly decide their life chance is enhanced by staying and finding a good winter territory before their competitors get back from breeding.

Making a good decision at this point should increase their lifetime breeding success, because as we will see in the next chapter these birds have a challenging northwards journey.

4 NORTHWARDS MIGRATION

As a broad generalisation the migrations of spotted sandpipers and common sandpipers are spread over many weeks, and for both species the early arrivals on breeding grounds are before some birds have left the non-breeding areas in the tropics.

GEOLOCATOR INFORMATION FOR BREEDING COMMON SANDPIPERS FROM SCOTLAND

The most complete itinerary is provided by a geolocator (Figure 29), and it was surprising how late this common sandpiper from West Africa left (Bates et al 2012), and how rapidly it progressed. Between late February and the first week in April it moved up to the southern edge of the Sahara, and left on April 10 to cross the desert in a day. At ground level there were strong north-easterly winds, but by getting up to 2,000–3,000 metres it found the wind coming from a totally different direction, aiding a northerly migration speed of up to 45 km per hour.

The route it appears to have taken across the desert approximates very closely to the route of a major pre-desert river (Coulthard et al 2013), and along the south side of the Atlas Mountains there are oases and seasonal streams as potential refuges if something were to go wrong. (This route appears to be taken by other birds that have had radio tags fitted, like cuckoos that have wintered in central Africa, seen to move west to then turn north along a similar route, along with nightingale, ospreys and honey buzzards.) The sandpiper spent a week somewhere close to the short sea-crossing from Morocco to Spain. During this period there was some cold and windy weather, with Almeria reporting gusts of 98 km per hour. This generally westerly weather possibly led to its movement to the Costa Brava (several later birds' geolocators went directly north, with stopovers in central Iberia). It then moved on rapidly to the narrow part of the Channel, then to the same areas in north-west England and central Scotland where it had stopped on the way south. During this movement the wind was easterly, so whether these places were visited by intent (known from previous route) or weather-induced is debatable, as a straight line would take it up the east coast.

This bird has left the Sahel at the end of the dry season, but I would guess it left the coastal mangroves in good fettle and was able to await somewhere around a big river like the Niger or the Senegal for appropriate weather conditions, to set off across the desert.

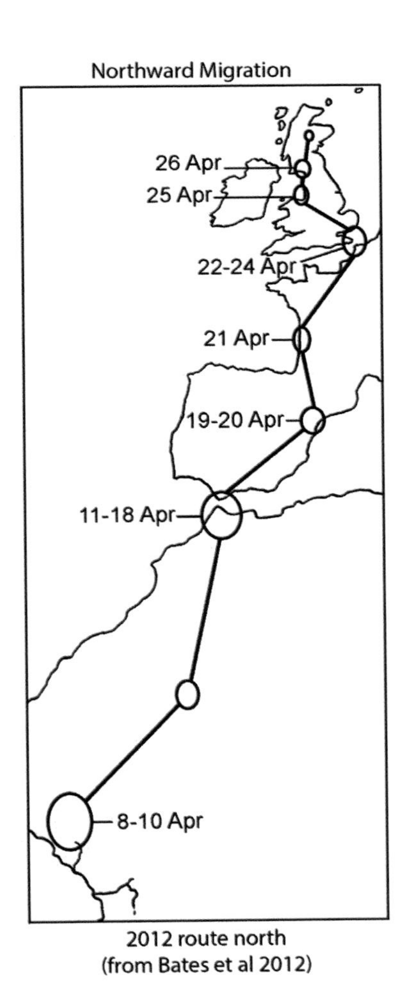

Figure 29: The route and schedule of the first common sandpiper fitted with a geolocator (Bates et al 2012).

Although the general area where it paused in early April (the Massif de l'Affole) appears at first sight to be desert, there are actually some wet areas around there, with a small population of crocodiles.

More Scottish breeders have had geolocators fitted since, and show variations on the same general route with some staying longer in Spain (Brian Bates, priv. comm.). The interactions with the weather show the importance of flexibility and there being a variety of places that individuals can stop to rest and top up with food. In geolocator-fitted individuals, the northwards migration is usually to the east of the southwards. Their arrival times across the south coast of England varied between 12 April and 4 May (they had all been breeding adults in their previous year).

On a family holiday on Fuerteventura in the Canary Islands from 30 March to 13 April 1994, I first noticed a common sandpiper on 1 April on a puddle caused by a water leak just outside the hotel. Over the remaining 12 days there were always birds there, and on the rocky seashore pools nearby; and on inland walks wherever there was water there was at least one. On the 8 April two were displaying at the puddle for five minutes, one erect and pushing its head upwards and forwards while the other turned its back and fanned its tail, keeping its head down. Reports on migration across the desert mention regular stops by a few sandpipers (e.g. Smith 1968 in south-east Morocco, where 20 arrived together when wind was in the north on 28 April, and Cramp & Conder 1970 at an oasis in the centre of Libya, the birds seen daily while the observers were there in the first week of April).

AFTER CROSSING THE SAHARA AND THE MEDITERRANEAN SEA

Birds crossing the Mediterranean Sea in spring have been caught on the island of Montecristo off the Italian coast, halfway between Tuscany and Corsica (Bacetti et al 1992). The birds pausing there were caught between 11 April and 20 May in the years 1988 to 1991 in mist nets on a small north-west facing bay, which was the only habitat suitable for waders on the rocky island. At any one time the maximum number of birds was three, and on roughly half of the days there were none. The distribution of weights is shown in Figure 30. Five of the birds were trapped again after a few days, and one had put on 1.25 grams per day over six days (37.5–45 grams); a good demonstration of how they can find food on what looks unpromising ground. The lowest weight of 28 grams is extremely low, though the authors also quote one of 25 grams on Malta. Although those particular birds were not observed to be in distress, it is clear from the data in earlier chapters (where weights on breeding, southwards migration and wintering are rarely below 40 grams) that most of these birds needed rest and recuperation. One of the birds was later trapped in northern France, so it survived the crossing. Seven individuals were caught 170 km further north on mainland Italy, and their range was 37–46 grams, so all of those were heavier than the mean at first capture (36.4 grams) of the island birds. Some caught on Defilia oasis in south-east Morocco were 35–43 grams (Ash 1969).

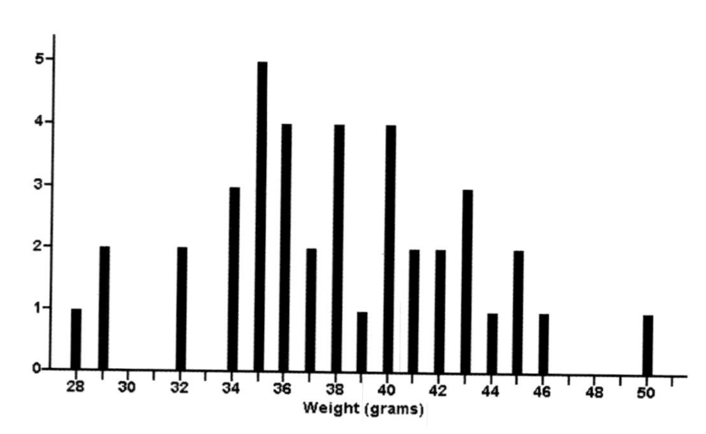

Figure 30: The histogram of weights of birds landing on an island in the Mediterranean; these are much lower than those in Table 2.2.

Bannerman reports that birds 'swarm through Gibraltar in groups of 4 or 5 peaking on 15 April'. Nearly everywhere through France and Germany there are birds from early April to late May, but if a site has a big peak of passage, it is in early May (OAG Munster 1982). The median arrival time in north Norway is 20 May.

As the birds are in such a hurry and most bird ringers are getting down to serious breeding studies, the number caught, measured and ringed on the northwards passage is much lower than the southwards, and thus rarely enough to publish. Nonetheless Table 4.1 below summarises limited data:

Table 4.1

Place	Latitude	Dates	Mean Weight (and range)	Comment
Italy	42°N	11.4–20.5	39g (28–54g)	after crossing Sahara and Mediterranean
SE England	51°N	20.4–12.5	53g (52–59g)	pausing
Norway, breeding	63°N	late May	54g (42–65g)	settled (Lofaldi 1981)

In contrast to the southwards migration, no heavy birds have to my knowledge been reported on northwards migration through Europe, when they feed on the go as they need. Furthermore there is little evidence of regular stopovers: birds caught in spring are not retrapped or re-seen at the same place in later years. The numbers in one place are erratic from year to year; I suspect that this is generally true for all inland wader migration. Akriotis 1991 showed wood sandpipers in southern Greece, stopping for short times after crossing the sea and putting on a little weight (circa 5 grams over three days) and moving on. One bird stopped for two weeks but only added 1 gram, so was probably sick. No correlation was found with the amount of water, ages of birds, or sizes of birds. The peak date was at the end of April. It appeared that birds dropped in for rest and recuperation at random.

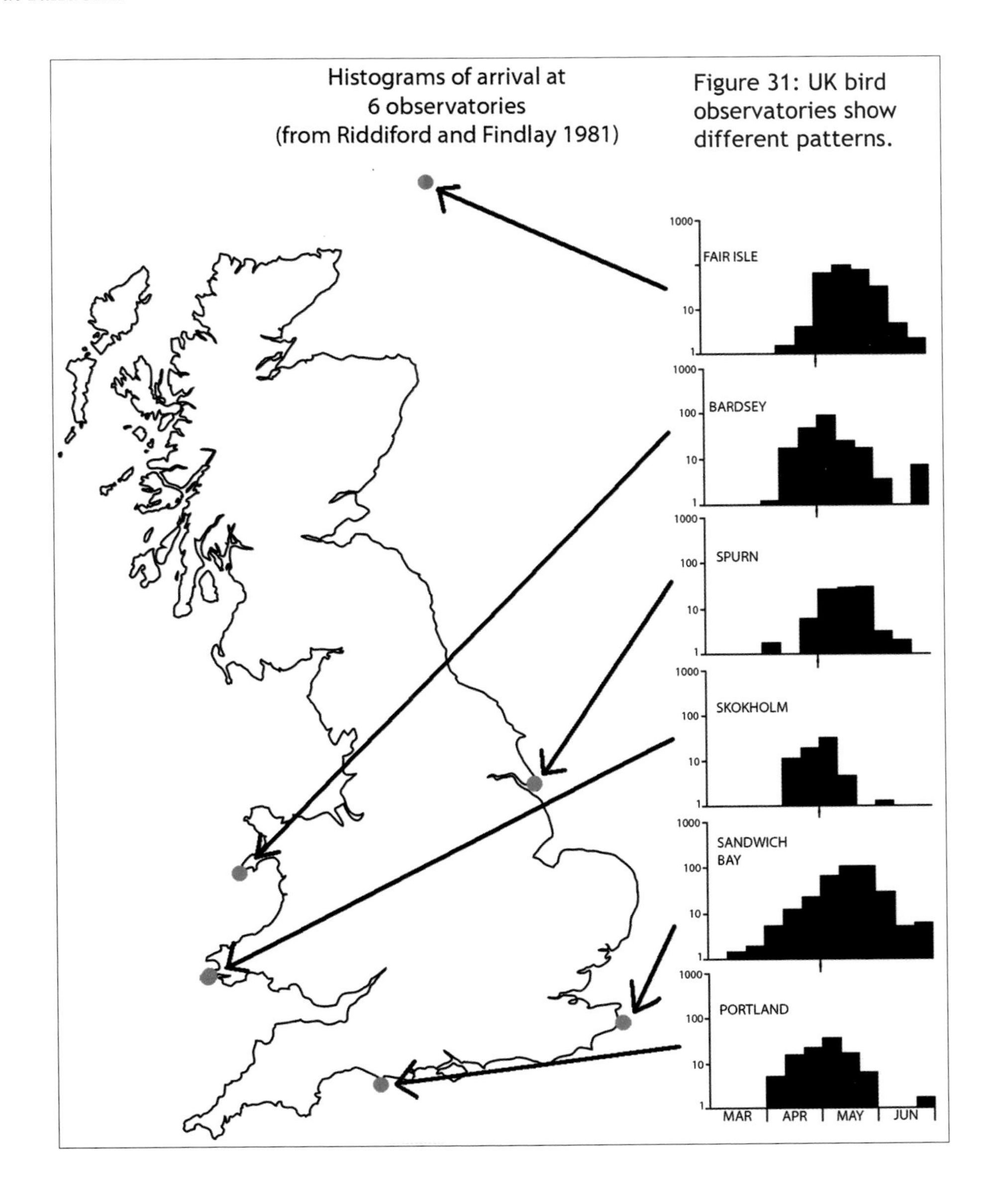

Figure 31: UK bird observatories show different patterns.

At British bird observatories the patterns (Figure 31) are fairly distinct from one another: those on the west coast have smaller numbers and the peaks are associated with the British timetable of breeding, while on the east coast the arrivals are later and mainly heading for Norway. The observatory on the south-east tip at Sandwich Bay has the longest season, and is the only one with March passage: this could be birds that have wintered in Europe, starting a slow movement north and eastwards. The numbers on Fair Isle in spring are as high as anywhere, which at first sight is surprising because the only place they only breed further north is in Norway. But as the sea-crossing to Norway from Fair Isle is half that direct from England (Norfolk) it is reasonable that the route from Spain via Brittany and Scotland to Norway is used.

Very few individuals are found dead in the British Isles, but from the study populations there have been an apparently disproportionate number in April:

Table 4.2 Recoveries in the breeding season are extremely rare; most are in the northwards/arrival weeks

14 April 1979	Combs, Derbyshire	Peak District adult from 1973
22 April 2004	Devizes, Wiltshire	Borders chick from 1996
26 April 1981	Peak District during extreme snow event	Peak District adult from 1977
30 April 2004	Ingram, Northumberland	Borders chick from 2001

The last was in a traffic collision, so it is possible that it had actually bred thereabouts in the previous two years and was not stressed by migration but just unlucky.

These were all adult birds at the time of death, so my inference is that they were trying too hard to get back early to their breeding territory. Up to 1991 more than half the dead recoveries in Britain were from the arrival period to mid-May (Holland & Yalden 1991b).

It is unclear how much stopping is done just for a rest, or how much feeding up they do, or how much is assessing the place as a potential breeding site. Probably all those reasons may apply, but in varying amounts for each individual and each place. As an example, during the last two weeks of April when information from geolocators and common sense would say that thousands of experienced British breeders are crossing Sussex each day (which has 150 km of coastline and lots of habitat and lots of birdwatchers out in springtime looking for migrants), only ones and twos are reported, because nearly all are flying over at night, focusing on getting to their breeding area. In May the numbers that are seen increase, as those needing a rest before going on to Norway are coming through. Those few that have been caught are no different in weight than those caught on the breeding ground and are much heavier than those caught on the Italian coast, so feeding is a choice rather than a necessity. One I caught in Sussex was found breeding in the north of Scotland a fortnight later (Figure 1) so that bird appears to have just been having a rest. One that Derek Yalden caught by Ladybower reservoir was seen again the next day 150 km north, and was possibly assessing the Peak District; being handled by a large mammal may have led to it deciding to move on!

FROM SOUTHERN AFRICA AND ELSEWHERE

These birds have further to go to reach their breeding sites, and thus rather than leaving in the second week in April as seen for West Africa, the adults are reported as leaving from early March (Tree 2008). A bird in southern Africa has been recorded as heavy as 93 grams (thus even bigger than any bird in Europe starting its southbound journey). Some recoveries further north in Africa suggest they stop in Sudan (Underhill et al 1999), and I would guess that the Nile Valley is a useful place to rest and refuel as necessary on the way.

In a thorough study of spring migration through lagoons on the Black Sea coast of Romania from 1990 to 1996, where there were up to 50,000 waders with mean counts of 5,000 curlew sandpipers (*Calidris ferroginea*), 8,000 ruff (*Philomachus pugnax*) and 9,000 black-tailed godwits (*Limosa limosa*) the maximum count of common sandpipers was 30 in mid-April in one year (Schmitz et al 1999). So, again, stopping is unusual and random. Their arrival schedule in western Russia is fairly similar to that in the British Isles.

As we saw in Chapter 1, the arrival of spotted sandpipers in the breeding grounds in Minnesota is in the second half of May. Although it is at only 47°N, it has a continental climate, with lakes frozen into late April. To reinforce the view that this is a temperature-driven date rather than that species being inherently later, the arrival of common sandpipers in Mongolia at 48°N with a similar continental climate was also noted from mid-May (Kitson 1979).

IS GLOBAL WARMING CAUSING EARLIER ARRIVAL?

Analysis of a 26-year series of spring passage dates in Sweden (Adamik & Pietruszkova 2008) showed that the median date moved earlier, at 0.23 days per year around a median of 3 May. In north-east Scotland the arrival of the first bird in Aboyne, Aberdeenshire, was noted from 1974 to 2010 (Jenkins and Sparks 2010) and was found to be earlier by 0.32 days per year around a median of 22 April. This area of Scotland experienced a mean temperature rise (January to May average) of 1.4°C over the same period, which had an even greater effect on short-distance migrant birds like pied wagtails (*Motacilla alba*) and resulted in much earlier song from resident birds.

In a comparison of field notes between 1875 and 1884 and 1975 to 1984 from the Barnsley area, 25 km north-east of the Peak District, the first arrivals in the more recent period were found to average 25 days earlier (Lunn 1992). I suspect that is a consequence of increased bird-watching, and of improved habitat with mining subsidence producing nice springtime stopovers at 'flashes' in the area; at face value this is 0.25 days earlier per year over the 100 years when global warming was hardly happening.

Over the period from 1977 to present on our Peak District study area there has been no significant change. The birds arrive at the Scottish study streams on average before ours in the Ashop, regardless of ours being 300 km further south, so I get the impression of birds in the British Isles appearing all over the place, as if raining from the sky onto attractive places, rather than any wave-front steadily moving north.

The date of arrival in Manchester in 1814–21 was quoted as 29 April (Blackwall 1822) which is no different from the general date of arrival in the area now.

Overall there is a variable picture, but possibly a generally earlier arrival over the last few decades of warming.

SPOTTED SANDPIPER

BNA summarises the situation: birds have left Paraguay by the end of March, and Peru and Brazil by 25 April. In Suriname they start to leave in March but the last does not go until June. They are passing through Venezuela and Costa Rica through April and May. The peak in Florida is the third week of April. The arrival in Minnesota has been covered in Chapter 1, but is essentially from the middle of May. Again the first breeders are back on the breeding grounds before the latest leavers have left the tropics.

They have been moving up the west coast to be in Oregon during late April, in south British Columbia in mid-May and in Alaska in late May.

More birds appear to go up through the centre of the continent on northwards migration than southwards. At Squaw Creek National Wildlife Refuge in Missouri in 2003–04 there were 282 counted in spring but only 25 in fall (Farmer & Durbian 2006). BNA describes a three-year dataset from Norman, Oklahoma, where only 9 birds were recorded on southwards passage (12 July–12 Sept) whereas 68 were seen northwards (19 April–31 May). Obviously these are tiny numbers compared with the continental population, but are a sample of those flying over in the same way that any site in Europe is only seeing a tiny fraction of those flying over. There is far more wetland up the centre in spring than in fall, and the general weather assistance will be to start westward along the coast from Brazil towards Venezuela and pick up southerly winds (Figure 32).

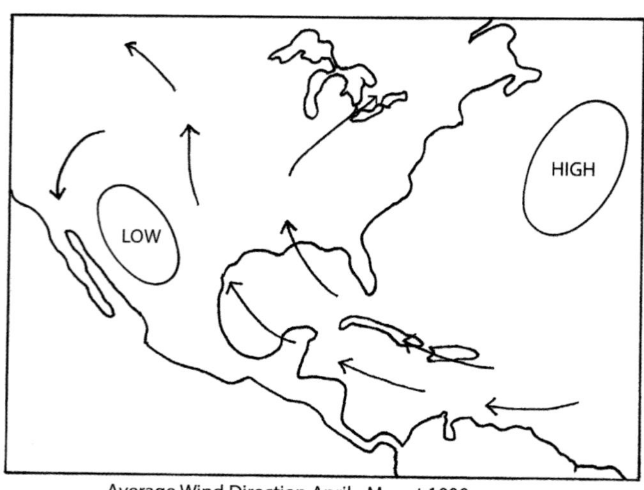

Figure 32: The average synoptic situation in spring supports spotted sandpipers moving west into the Caribbean and then going up the centre of the USA.

SELECTION PRESSURE

We have seen that once a common sandpiper chick has become strong enough to fly away from its breeding place, it then joins the procession of common sandpipers moving south during a time of plentiful food and generally favourable weather. Although some sickly or unlucky individuals might die, there is not a desperate struggle for survival. Once those heading across Africa reach the Sahara they need to have built up enough reserves for the crossing, but there is no rush. In West African mangroves there is competition for territory, and again the weakest will be pushed into marginal habitats. But probably the main test of fitness is judging the time and route for this northward migration, when the dryness of the Sahel reduces places to feed and when the general weather at ground level is discouraging; then when they reach Europe the weather is extremely variable. Negotiating this obstacle course is a good test of their total fitness, and if they arrive back on the breeding areas they have proved their basic capability. For a first-year bird the intelligent thing is to survive rather than to rush back, because it is not likely to get the best territory or mate. The key task is to get back safely and look for a good place, and to gain a first territory without too much of a fight.

Sexual selection then comes into play, and the female chooses who to breed with. If she finds the male that she knows from a previous year is already there he is obviously still fairly fit, and if he exhibits good aerobatics and defends his territory he is definitely fit. If she finds he is not there, she looks for another who has arrived and is doing his stuff. If she is a first-year female there may be an experienced male waiting, or else she will go with a new one.

Spotted sandpipers appear to have the same basic tests of fitness but on arrival at the breeding area the male finds that the female has already staked out an area she wants to breed in, and her first choice is between the early-arriving males (all of whom have thus met the basic fitness criterion). It is clear that some males have to wait in line. Probably that male which is first for her eggs is one that she recognises from an earlier year, but other aspects of their displaying flights and wing-raising postures will help her decision making. One thing she does know is that a decision needs making quickly, and even if her first choice is not optimum she will have another choice a week later.

Return to Chapter 1 to find breeding behaviour from arrival: jump to Chapter 7 to see how populations are changing, or go to the next chapter to see what they eat.

5 FEEDING

Some reference to food has been necessary in earlier chapters as we have reviewed the sandpipers' life cycle, but in this chapter we look more systematically at how they feed and what they eat, and how much they need to eat so they can do what they need to do.

There are various ways of assessing their diet. The starting point is watching how and where they feed, and during such observations they will occasionally take an identifiable item. In earlier days when ornithology and shooting were closely related the analysis of stomach contents of killed birds provided further information. Then in an age of ecology the sampling of the available food and the study of faeces and pellets to check which remnants of prey were discernible therein added further information. The literature is full of observations on their feeding; and 'Common Sandpiper Feeding on Hippopotamus?' (Hassell 2006) is thought-provoking.

Both common and spotted sandpipers' bills are straight, and their length is similar to their head depth. The length allows them to extract food from between stones or stalks where it may be lurking, but the bill is rarely used for probing into mud. One afternoon I crawled around a breeding territory with a graduated stick the thickness of the bill, seeing how many places between pebbles or roots where an invertebrate might hide were less than 25 mm deep (and more than 10 mm) and found that 80 per cent of those I tried fitted within those parameters. Sandpipers' bill ends can feel, and the birds have eyes of a size that is normal for a bird that feeds by sight during daylight and align with the bill tip. Many shorebird species are similar, and indeed all those that have the vernacular name of sandpiper or stint have very similar bills except the spoon-billed and curlew sandpipers, highlighted as extraordinary by their name.

Much feeding activity occurs while the birds are walking close to the water's edge and pecking frequently (typically every one or two seconds). When feeding on grassland, its method is sometimes more similar to that of a thrush; looking ahead, then running and grabbing. Occasionally one appears to be stalking the water edge in a heron-like way. Other occasional modes are to fly and grab a flying insect, or to swim like a phalarope pecking off the surface. They do not feed in flocks of either their own kind or other shorebirds.

In the breeding season they feed on land or by fresh water, but on passage and in winter many feed in salt-water environments and so they have appropriate nasal glands to take salt out of the blood and secrete it (often called salt glands). On some breeding

Figure 33: A spotted sandpiper on passage through Texas, photographed by Greg Lavalty. This is about as big an item as they eat.

streams that exit direct onto the seashore, this gives them extra flexibility to feed there when useful. The mass of the nasal glands as a percentage of body mass at 0.06 per cent is similar to those of guillemots (*Uria aalge*), and half those of knots, but much greater than those of totally freshwater waders like green sandpipers (*Tringa ochropus*) at only 0.01 per cent (Staaland 1967). The sandpipers' glands are positioned just above the eyes, and the secretion comes out of the nostrils at essentially the concentration that is in seawater. However, they do not like extreme conditions, and unsurprisingly are reported as scarce on soda lakes in the African rift valleys and on passage in sodic-alkaline pans (e.g. Boros et al 2006).

THEIR FOOD

In this section, the common vernacular names of the sandpipers' prey are mostly used. In Appendix 2 is a list of scientific names to whatever taxonomic level they have been identified (rarely to species level), the main season they are eaten, the places where they are eaten and the source reference of the information is given. The appendix also says whether the source reference applies to common or spotted sandpiper, though in practice there appears to be a large degree of commonality in the food taken by the two species. Indeed the appendix is by nature incomplete, because it largely reflects food at the places where studies have been done, whereas over the whole world they will eat whatever is available and looks and behaves like the range of prey items sketched in Figure 34.

LARGE ITEMS VISIBLE TO HUMAN OBSERVERS

Much of the time when we are watching sandpipers feeding, they are walking along pecking away at tiny things, but occasionally they take an identifiable prey item which it takes time to eat, and sometimes the prey is taken to the water and washed before being consumed. Earthworms are taken from damp grassland, especially early in the breeding season. If caterpillars are descending from a tree they are a gift, and resting moths may be grabbed. In Minnesota when breezes blow myriads of insects onto shore, they are taken. At the edge of the water, where sandpipers are by reputation supposed to feed most of the time, they rarely take items that are identifiable, though on migration in saltmarsh or brackish river channels they may be seen occasionally with a ragworm or small crab.

Stanton 2013 observed a small gecko being eaten in Hong Kong. 'The bird manoeuvred the gecko so the head was hanging downwards, by tossing and catching it in its bill, before striking its head against the rocks and killing it. It then removed the tail which was eaten before the body was swallowed head first'. It was an immature gecko, about 30 mm long. Other unusual items that people have recorded include fish and frogs, and maybe these are commonly eaten when small enough ones are available; in a shallow riffle at 1,740 metres in the Swiss Alps, Burkli 1990 observed a juvenile eating a small fish 40–60 mm long (i.e. twice as long as its bill).

In mangroves in the non-breeding season, small crabs are visibly taken. Zwarts 1985 showed that sandpipers would feed on juvenile crabs with carapaces of around 5 mm (a full-grown one being 25 mm is far too large for a sandpiper, and one of 10 mm was observed to be rejected), and points out that a crustacean *Callicherus balssi* was very numerous along that channel and probably formed the most numerous prey. The means of catching crabs is to dart at them and grab them before they can get to their hole, which is far too deep for a sandpiper bill to reach down; getting them from deep down is a job for whimbrels. The size of 5 mm is essentially the gape of a common sandpiper.

Spotted sandpipers in Suriname were also observed feeding on fiddler crabs (Spaans 1979).

ITEMS THAT ARE DOMINANT IN THE AREA SANDPIPERS ARE PECKING, AND REASON SAYS THAT THEY ARE BEING EATEN

For spotted sandpipers, M&O say

> Our observations clearly indicated that terrestrial arthropods (primarily insects) were their main food source on Little Pelican Island. Although small aquatic organisms certainly were taken, it is unlikely that these constituted a substantial portion of the diet, with the possible exception of those periods when insects were very scarce. Most of the time these birds spent foraging along the shoreline they actually were consuming terrestrial insects washed in by wave action.

In my observations of spring passage where birds feed along a concrete dam (Figure 35) they must be picking mainly the small insects that the wind and waves bring in and

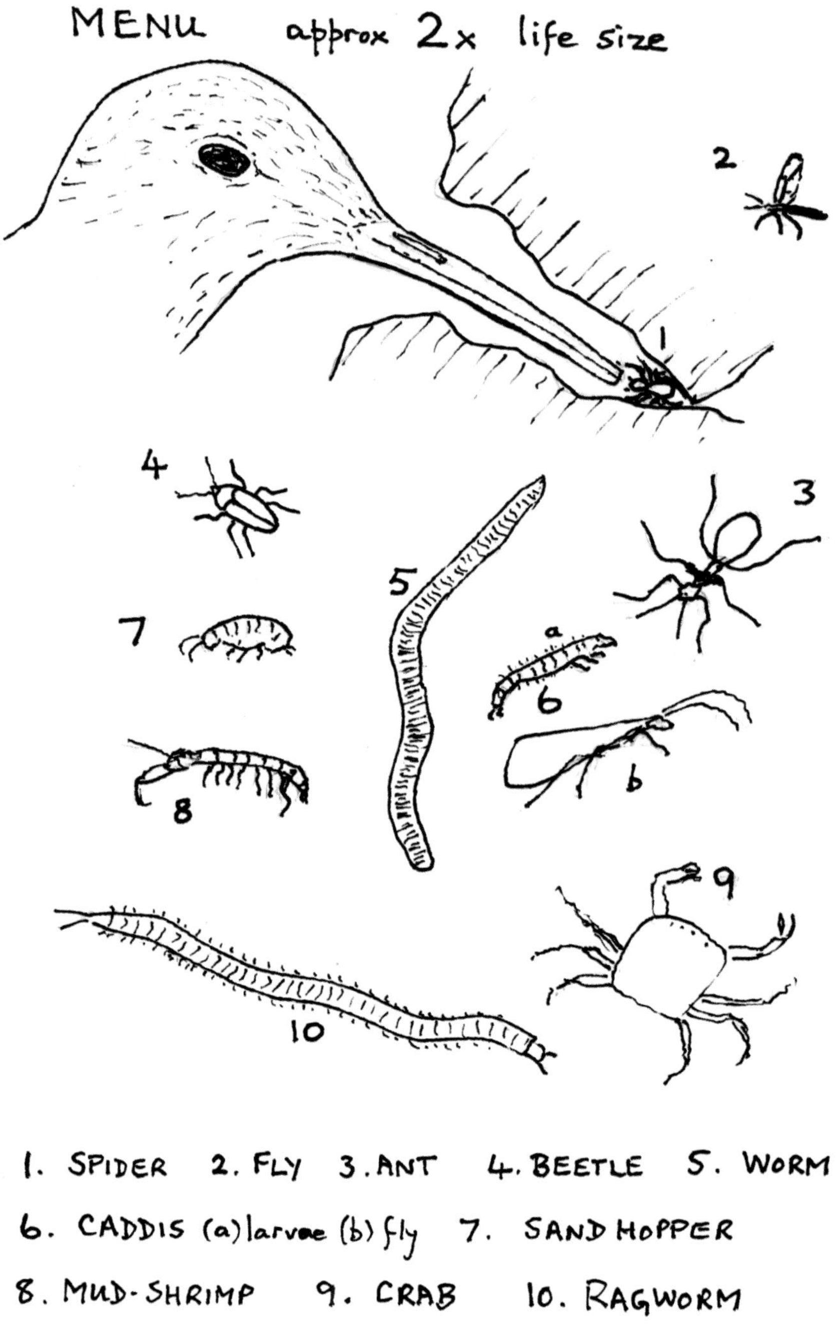

Figure 34: These are my sketches of typical things eaten by *Actitis;* anything of these basic sizes and shapes is potential food.

Figure 35: Concrete is not an obvious feeding substrate for a wader, but Weirwood dam in the south of England regularly has sandpipers on migration feeding on items blown or splashed onto the concrete, and spiders and insects living on flotsam and lurking in cracks.

which are only visible to a human crawling along, identifiable only by being picked up on sticky tape and with some magnification. But spiders and other creatures that live in the cracks in concrete must also be taken, because occasionally the birds will rush up from the water's edge towards the crest of the dam to grab something that it has seen. In my observation of late summer passage in saltmarsh they are pecking away rapidly at the top layer of stony mud full of the mud-shrimp (*Corophium volutator*), and so I presume that is their main food at that time.

In Yellowstone, Kuenzel & Wiegert 1973 recorded the pecking activity (quoting 3,638 pecks per day) on wooden platforms in hot springs where the algal mat was eaten by brine fly larvae and emerging brine flies. One bird was then checked, and it contained 98 fly larvae, 1 pupa and 2 adults, plus a spider and a beetle.

ITEMS RECORDED FROM STOMACHS

Killing birds to see what they eat is no longer a regular practice, so the data summarised in the standard books like BWP are still the main source of information. These suggest a flexible diet in which the birds are taking what is abundant and catchable; mainly craneflies in one sample, mainly mosquitoes in another, mainly beetles in another, but other types being taken as well – flies, spiders, ants, midges, grasshoppers, stoneflies, mayflies, bugs, crustaceans, earwigs, fish eggs. Seeds get mentioned, but the guess is that these are accidentally taken when invertebrates are grabbed.

Yalden 1986a (Y86a) used some accidental casualties (two adults, one juvenile, four chicks) and found the same items as deduced from observation and from analysis of faeces and pellets.

In Puerto Rico, mole crickets were found in spotted sandpiper stomachs (Wetmore 1916). This was a study to find which birds were beneficial in eating agricultural pests; nine stomachs were analysed during most months between the birds' arrival in July and departure in May. But the largest proportions of items (80 per cent) were crustaceans – fiddler crabs and amphipods – consistent with the sampling sites being near mangroves, and thus the same food as that eaten in Suriname and by common sandpipers in West Africa.

ITEMS FOUND BY FAECES ANALYSIS

Y86a reported on the remains detected in 94 faecal samples collected on breeding territories in the Ashop study area (Appendix 1). The samples were examined multiple times under different microscopes to ensure complete scanning, and items like wings and tibiae were also measured. Reference items were available from the locality by shuffle, pitfall and sweep-net sampling (more on these methods below). There were 553 prey occurrences; 1,362 individual prey items were identified, but for any one occurrence of a worm in a sample there could be 20 items. The crucial result was that 53 per cent of prey was from terrestrial groups and 20 per cent aquatic with the balance being equivocal, and this was consistent with the amount of time spent feeding in those habitats. A significant difference between chicks from 5 to 30 days old was that they spent nearly all the time on the shingle, and so earthworms were much scarcer in their faeces than in those of adults. Proctotrupoidea (small parasites of other insects) were more frequent in chicks. The earthworm difference is mainly due to time of year, as adults mainly eat them in April and early May when they feed often in fields. However at all ages, mainly beetles and weevils were taken. The weevils were interesting as they were not caught in shuffle sampling or pitfalls, and the guess is they fall into the stream from trees upstream. As we have seen earlier, particularly from studies of spotted sandpipers, items blown or swept by the water flow are important.

Figure 36: While a brood of chicks were being ringed, they were kept in a clean container and the faeces collected for microscopic detection of remains of the items they had eaten (Y86a).

Derek Yalden continued his interest in feeding by having a series of students doing projects from 1997 to 1999. Julien Martin looked at faeces from birds at Ladybower reservoir (into which the River Ashop runs). The range of food eaten was the same, but insects hatching from the water (mayfly, stonefly, caddis) were even less on the reservoir, at below 2 per cent. Up the River Ashop, ants were a significant part of the chick diet, while they were totally absent from Ladybower chicks. The adults at Ladybower, however, ate more ants, though a different species, with the woodland providing them with wood ants (*Formica rufa*) which may be too aggressive for chicks compared with the red ants (*Myrmica*) which they eat up the river valley. The Ladybower chicks also ate more caterpillars and weevils, consistent with the greater number of trees around the reservoir. Most of the trees are coniferous with acidic soil so there are few earthworms; only one earthworm was found in faeces. This was followed by further faecal sampling in 1998 by Elsie Ashworth and a more statistical review by Anita Hellstrom in 1999, giving the same broad result that the main food is terrestrial and they eat anything that is available. There was no significant difference between samples collected on the river in 1998 and those collected in 1979–84 reported in Yalden 1986a.

ITEMS MAINLY FOUND FROM PELLETS

Arcas studying in autumn on the Ria de Vigo in northwest Spain analysed 56 pellets (giving 610 prey items) as well as 32 faecal samples (giving 64 prey items). The most abundant items were isopods (present in 80 per cent of pellets and 90 per cent of faeces). Only in pellets were small snails, small crabs and insects detectable. Only in faecal samples were chironomids and beetle larvae detectable (Arcas 2000). In the winter he found that insects essentially disappeared, and nearly all the food items were then marine invertebrates, sandhoppers and ragworms making up 86 per cent (Arcas 2004).

Generally few items have been detected solely in pellets, as these consist of the hard parts of the creatures for which other evidence is normally found in faeces.

In Derek Yalden's correspondence is a letter from Eric Smith, an insect specialist who had found three pellets in our highest territory (at 300 metres) on 11 July 1982, and in these he identified the following genus of beetles *Agabus, Helophorus, Anacaena, Byrrhus, Nebria, Pterostichus, Haliplus, Omaliinae, Aphodius, Phyllobius, Hydroporus*, and also the earwig *Velia caprai*, which was in all three of the pellets. He comments on the total absence of caddis remains. Caddis larvae are abundant in the river and formed the major component of dipper pellets on the same length of stream. Although observation shows that dippers and common sandpipers use different feeding techniques, this clear difference in diet on a shared stretch of stream is nice to see. The few territories where both species occur tend to be rather unsuitable for common sandpipers, and indeed this high territory had not been used for the last 20 years.

GRAZING BIO-FILM

In the far east of Asia and far west of North America, a study of small sandpipers showed that they had special adaptations that enabled them to sweep up the nutrient from films

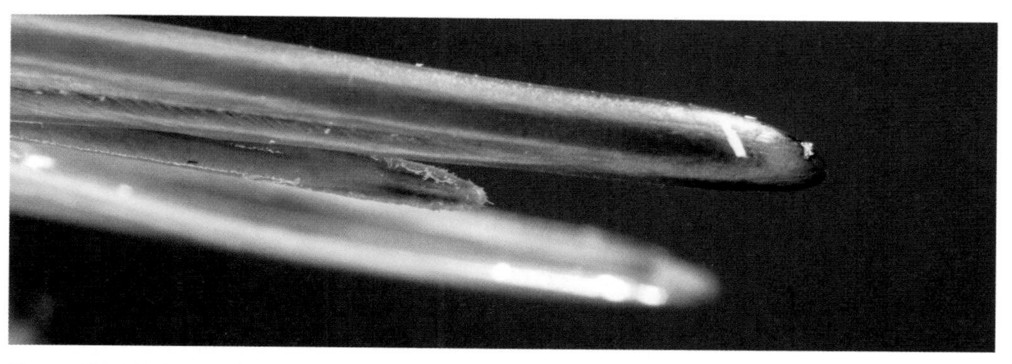

Figure 37: Birds use their tongue to manipulate food, but in this picture sent to me by Kuwae from Japan the bristles on the tongue indicate the capability he found for shorebirds to graze on biofilm.

of algae on the mud, and this was particularly shown for western sandpipers (*Calidris mauri*), finding roughly half their food intake by this mechanism (Kuwae et al 2008). Kuwae sent me a picture (Figure 37) of the tongue of a common sandpiper in Japan, with similar bristles to those found in grazers, suggesting that it would be capable of such a feeding mode though its prevalence is unknown.

AN INTERESTING EXPERIMENT

To see if sandpipers could be trained to probe, a non-breeder was caught in Thailand and kept in an enclosure, and the first tests were to see which food it preferred, from: pellets used as feed for shrimp farms (first choice), pieces of shrimp, live polychaete worms, and pellets used as chicken feed (last). They then tried with favoured food left visible or covered by a little sand, and the bird preferred that which was visible. They then tried to see if it could distinguish between clean sand and shrimp-flavoured sand. They found 32 probes in the shrimpy versus 6 in the clean. But as the test progressed the difference gradually reduced (perhaps because it never found food in either case!) and the bird eventually gave up. The bird was released and watched feeding in the local saltmarsh as normal (Swennen & Saeho 1993). This does suggest that they have a sense of taste/smell to help them to detect food.

THE TIMES OF ABUNDANCE OF KEY FOOD

We have seen that the sandpiper diet is very broad. Now we look at the abundance through the seasons of some key items to see how much the migratory pattern of the sandpipers enables them to go from one place of plenty to the next.

BREEDING GROUNDS

Yalden 1986a gave results of various forms of sampling. Shuffle samples are taken by shuffling among the stones in the riverbed for a given time while holding a net to catch what has been dislodged. Pitfalls are containers sunk into the river bank to sample what

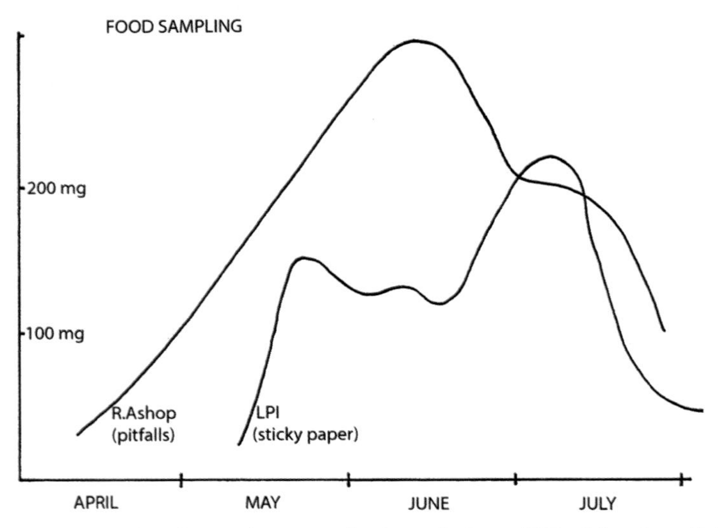

Figure 38: This shows the smoothed graph from BNA of food availability for spotted sandpipers on LPI (1974-82) as measured by sticky cylinders catching flying insects, and that for food sampling in the Ashop as measured in pitfall traps to see seasonal variation of terrestrial prey for common sandpipers (Yalden 1986a Figure 2, average of 1982 and 1983). The increase in prey happens at the time that the birds arrive (three weeks earlier in Ashop than LPI). The differences do not represent different requirements of the two species; just the different kinds of habitats where studies were done: an island on a lake that is frozen in winter suddenly gets a burst of insects at thaw, the first of those that have spent their larval stage under the ice hatch. The food in a stream that rarely freezes and where much of the birds' prey is beetles increases more sedately. The y-axis value for the two sampling methods happens to be in the same general range, but this does not mean that there is more food by the stream: it is the shape that matters, with the rapid rise in spring and the reduction by late July. The second peak in the LPI food availability was of note, as the percentage of clutches that successfully hatched on LPI was higher as June progressed. Not shown in the picture are the times of hatching stoneflies and mayflies in the Ashop, which can give added bursts of food.

is running about; if the water level rises too much all they catch is water! In one sense the data just give a scientific face to what is clear to anyone who sits by the places where sandpipers breed: in mid-April one may see an occasional insect or spider on the bank, the trees have no leaves, and river water is cold and appears fairly lifeless. On into May and especially on a warm day the banks become active with flowers, terrestrial insects, spiders and other insects that start to emerge from the water. By the end of June most creatures that had been larvae in the water have emerged, providing the flush of food that the feeding young birds need, and so the shuffle samples drop. The peaks of emerging aquatic prey and shoreline life in late June match the main time when chicks are growing. Many items of food then either aestivate or fly off to lay eggs as July moves into August so food that is catchable becomes scarcer.

M&O showed that there were two peaks of abundance at LPI, and although these were different in time and relative size from year to year they were always about four weeks apart. The first peak was midges, and the second midges plus mayflies, with the mayflies being less in number but of greater mass. In Minnesota, the sampling was done with sticky cylinders that caught flying insects. These were weak flyers whose distribution depended on the wind. The number of mayflies was such that 'at times freshly dead or dying mayflies carpeted the forest floor and accumulated in piles up to several cm deep along the entire length of windward beaches'.

ON PASSAGE

For birds leaving north-west Europe with opportunities for feeding on estuaries, the prey that is later exploited by flocks of Arctic waders is reaching its peak during the warmth of summer. As a textbook example the production of the mud shrimp (*Corophium volutator*) is increasing strongly through June, July and August, and these (as well as being food for sandpipers) are food for the young crabs and shrimps that themselves can also be food for sandpipers (and other passage waders). The range of prey items that sandpipers eat means that something is available in plenty through July to September. Similar peaks of food occur on inland waters.

NON-BREEDING AREAS

In the tropics north of the equator where sandpipers live from August to March, the food in mangroves that they can feed on is abundant: the fiddler crabs are hatching because their food source is good, and the young crabs are also food. Most waders feed out on the tidal mudflats, and the *Actitis* sandpipers are the commonest small waders feeding among the roots and mud of the mangroves. Those that are excluded from mangroves use their flexibility, which enables them to follow rain-filled pools, to use rivers and to exploit man-made wet places like sewage treatment sites.

Figure 39: The footprints of a sandpiper feeding in winter on the tidal mud of the River Ouse in Sussex near the water but not at the water's edge.

ENERGY USED, AND HOW MUCH FOOD IS NEEDED FOR VARIOUS ACTIVITIES

Birds do not move from continent to continent just to enjoy the view. Food is a major driver, but so also is avoiding extreme weather. Some activities are clearly much more demanding than others, most notably laying lots of large eggs, with the spotted sandpiper sometimes producing four times her own body mass as eggs. The energy used is estimated in various ways. A foundation of energy use is the Basal Metabolic Rate (BMR) which is that needed just to keep the body ticking over. This can be found in a captive bird that does not need to expend any excess energy and whose output of carbon dioxide can be measured allowing its rate of fuel burn to be estimated. This has been done on many species of bird by many different research groups, and enables a good estimate for any other to be interpolated. To measure energy use in a free-living bird doubly labelled water can be used: this is water in which much of the hydrogen isotope H_1 is replaced by deuterium H_2 and the oxygen O_{16} replaced by O_{18}, which is injected into the bird and allowed to equilibrate in the blood for an hour and then be sampled. The bird is then left to do what it wants for a day, and is then caught again and the blood sampled. The reduction in labelled water is related to the rate of energy usage and has been calibrated against the first method in captive birds.

The energy in eggs is directly measurable, and a general efficiency of egg production is used to estimate the energy expended by the bird. For the range of activities that a bird undertakes in a day there are assumed multiples of BMR that are used, though these tend to get updated as more is learned about avian biology.

BREEDING SEASON

Energy needs for various activities in the breeding season for spotted sandpipers, and how food supplies those needs were assessed thoroughly by M&O and is the basis for the following. The numbers are for a female spotted sandpiper. A male common sandpiper is much the same weight, so will be similar.

Basal Metabolic Rate BMR – 1.36 kJ per hour

During observations of activity, each was assigned a multiplier of BMR:

- roosting overnight 1×,
- foraging 3×
- preening 2.5×
- resting 1.24×
- flying 15.2×
- walking 2.8×
- courtship 3.5×
- nest building 3× and
- incubating 1.3×.

Territory defence was subdivided into:

- appeasement 1.5×
- standing upright 3.4×
- balanced response 4×
- chasing 5×and
- fighting 15.2×.

Most of the activities during daylight take about 3× BMR, and this multiple is commonly found in other species, so a typical Daily Energy Expenditure (DEE) for a female spotted sandpiper was calculated as around 65 kJ. One multiplier which modern data does not support is for incubation, which is now seen as much more costly, especially for a sandpiper clutch that has the same mass as the bird and is resting on the relatively cold ground. Direct measurements on breeding common sandpipers using doubly labelled water were reported by Tatner & Bryant 1993, and they found DEE to be 135 kJ during incubation. Allowing that it is a bigger bird, this is consistent with incubation for 12 hours overnight being over 2.5 × BMR. For Arctic breeding sandpipers the usual value taken nowadays is 4× (Deeming & Reynolds 2015). This recent uprating of incubation is not too important for a female spotted sandpiper, who does little incubation. Another of the multipliers that is now taken differently is flying where the usual value is about 9× (Piersma & van Gils 2011), but again, while on breeding grounds they do not spend much time flying. Fighting, for which the same very high multiplier of 15.2 was used, may well be exceptionally costly, and birds avoid it so it is only rarely that it is an energy-sapping activity.

Thus the DEE of around 65 kJ for a female spotted sandpiper not including egg production is still reasonable. (M&O give detailed breakdowns for several different scenarios of male and female stages in the breeding season but as that paper runs to 63 pages people who want that level of detail need to read it.)

MAKING EGGS

The average spotted sandpiper egg is 9.5 grams and contains 52 kJ, and the use of a standard 'efficiency' of 70 per cent means 74 kJ is needed for each egg. The process of growing and laying four eggs takes around eight days, so an extra 37 kJ per day is needed. She therefore needs to feed 1.6 times as fast during that stage, and as she does not feed at night that means feeding roughly twice as fast during daylight. She has another need, because her eggshells are high in calcium, and most of her food is not.

FOOD ON THE BREEDING GROUNDS

The calorie and calcium content of food was also studied in detail in M&O. For their midges, mayflies and amphipods, the energy value was about 22.7 kJ per gram (ash-free dry weight). Values for other typical prey are similar: thus Hale 1980 gives values for earthworms 17.3, craneflies 22.8, flies 24.4 and beetles 24.9. Clearly the number of individual items needed to make up 1 gram differs, from a few earthworms to many hundreds of midges.

On the breeding grounds of spotted sandpipers the importance of amphipods was that they contained 5 per cent calcium, roughly ten times more than the insects. For a species trying to lay big eggs and multiple clutches, calcium could be the limiting factor. (For five clutches a bird would need around 3.4 grams, which is much larger than the whole bird's skeleton calcium of approximately 0.4 grams, and even for one clutch it is clearly a potential constraint). For common sandpipers the insects may be adequate, but also their stopovers at estuaries on the northwards migration may give them some useful input by eating *Hydrobia* snails from the mud. Hale 1980 points out that *Hydrobia* size is at a maximum in spring, averaging 4.5 mg, and although very low on calories (7.8 kJ/g) they are high in calcium. When four species of Arctic sandpipers were sampled, it was found that 38 per cent of the stomachs of females had small bits of lemming bones, whereas only 2 per cent of males did, and it was postulated that this was a deliberate way of getting calcium (Underhill 1994). When a clutch is predated, birds may re-use the eggshells. Obviously they do get enough calcium somehow.

AN INDIVIDUAL FROM HATCHING TO HAVING ITS OWN CHICKS

Growing

The previous discussion has all been about the adult. But a few weeks into the season there are chicks that need to grow. The energy requirements for this have not been measured in our sandpipers but can be estimated from other waders (Schekkerman & Visser 2001). They studied black-tailed godwits and lapwings, and found the total metabolic energy was around a third higher than for bird species that were fed in the nest, because wader chicks have to walk about finding their own food and as they grow they need to keep themselves warm rather than cuddling their siblings in an insulated nest. Scaling from the godwit, which is about 6 times bigger and grows 6 times faster, gives a daily energy use over the sandpiper's 20 days from hatch to fledging of 70 kJ per day; much the same as the adult.

On a crowded site like LPI with polyandry, there is the potential for females still producing eggs while first chicks are growing and males are feeding, so it is not surprising that aggression to chicks was observed because even on LPI where food is abundant it is not infinite.

Migrating south

The fuel to get to the tropics can be taken on board fairly easily during the summer. A typical bird that will be going around 4,000 km (Chapter 2) will put on around 40 grams of extra weight, mainly as fat; and using the value of 38 kJ per gram (Newton 2008), the amount needed is around 1,520 kJ. Over about 20 days that is 76 kJ per day, thus doubling its necessary food input during that time. Using mud shrimps (*Corophium volutator*) as an example food on British estuaries, their mean weight in August is about 0.5 mg, containing around 5 joules, so while preparing for migration a bird, if solely reliant on the shrimps will need around 300,000 of them. Figures quoted for the density of *Corophium* vary widely with time and place; their accessibility disappears at high tide and the sandpipers do not normally feed at night, but values of 200,000 per square metre are reported (Adam 1990) though 20,000 may be more usual. On any reckoning they are very abundant, and at this stage the birds are not territorial. Even so, pecking at once per second, there would be 80 hours of extra pecking (plus associated time for digesting and ejecting waste) necessary to be able to migrate.

The obvious requirement is that birds find excellent food resources before they migrate in large hops, or else a steady sequence of good places to refuel between short hops. Both options seem to be widely available in the northern summer.

Moulting and surviving the winter

One of the key food items in mangroves is fiddler crabs (*Uca*), and in Guinea-Bissau further detailed observations on their distribution and behaviour was done by Ens (in Wolff 1998). Fiddler crabs were present along about 30 per cent of transects done. The occupied areas had about 0.737 grams per square metre. The density was much higher (up to 11 grams) in well-drained areas at the edge of channels. Sandpipers mainly feed up these channels. The crabs feed most in the heat of the day, and appear to have competition for space so that between looking out for competitor crabs, actually feeding and also watching for predatory sandpipers, the crabs have to be balanced in their time budget.

Moult takes some extra energy too. Feathers have an energy content of around 26.4 kJ per gram, so on the face of it that should not be too demanding for a bird with only around 2 grams of feathers. This is less than the daily energy expenditure or the laying of a single egg. Unfortunately the 'efficiency' of making feathers is much less than the efficiency of making eggs or exercising muscles, and for a bird the size of a sandpiper this efficiency seems to be only about 8 per cent (Hoye & Buttermer 2011). Even so, the finding of around 600 kJ over a few months should not be too difficult. A chick grows a full set of feathers in about 15 days, and the leisurely rate of moult of adults is probably driven more by the need to retain good flying capability and other factors rather than by energy. The decision of some first-years to only moult half of their primary feathers in southern Africa (Chapter 3) may be a response to a severe shortage of food in the ephemeral wetland habitat (Remisiewicz 2011), or it could just be that the timetable of these birds, which migrate more than twice as far as the average *Actitis*, leads to reduced moult being frequent, because they need to stop moult to focus on migrating.

Breeding

For the northwards migration, finding excellent feeding may be a greater challenge especially when crossing the Sahara and the Mediterranean. The birds caught when only weighing 30 grams (Chapter 4) must have used all their fat and started to deplete their protein from other tissues, which give much less energy per gram. As a wintering bird in a warm climate still weighs over 40 grams, these very light birds are fairly close to death. They need to be finding over 100 kJ per day to get back on the road.

SUMMARY

In this chapter I have jumbled up the sources of information relating to common and spotted quite recklessly, because there appears to be no real difference between the two species – though a small male spotted sandpiper will need less food and a large female common sandpiper will need more. The general trend for an individual is to need around 65 kJ per day. But for a spotted female laying her fourth clutch of eggs and for her earlier mates and their growing chicks, the multiple families' need goes up potentially to over 1,000 kJ per day, and the territory needs to provide it. The birds fly thousands of kilometres to procure a territory that is capable of providing it. One advantage of the male being slightly smaller than the female is that he needs slightly less food and can spend more time watching the chicks. Nevertheless, he obviously needs to be big enough to brood four large eggs, so being about 10 per cent smaller than her is a reasonable compromise.

6 PREDATION, COMPETITION AND OTHER NUISANCES

The previous chapters describing where and how sandpipers spend the year have largely ignored the other creatures with which they interact. In this chapter we will look at three types of interactions:

- Predators that clearly see the sandpiper eggs, chicks or adults as food
- Competitors that are using similar resources
- Other things that may affect their fitness.

PREDATORS

The study sites (described in Appendix 1), which have had good numbers of birds to observe, thus enabling plenty of data to be gathered in a reasonable time, have had some human control of predators. Thus in the early years of spotted sandpiper research on Little Pelican Island the deer mouse (*Peromyscus maniculatus*) caused significant egg loss, and mink (*Mustela vison*) were also a pest; control measures were taken. At the end of the study, mink, gulls and ruddy turnstones (*Arenaria interpres*) took a major role, and the last became objects of study. In the Peak District there is a sizeable interest to protect birds such as red grouse (*Lagopus lagopus scoticus*) for shooting, so that control of their predators is active. At the start of our studies in the Peak District in the 1970s, various birds of prey had been seriously reduced in numbers by pesticides. The main farming is sheep, so foxes are unwelcome. No control was done specifically to protect common sandpipers.

PREDATION ON ADULT BIRDS IN THE BREEDING AREA

Sandpipers are vulnerable to *Accipter* species (e.g. sparrowhawk, *A. nisus*, in Europe). The table gives some observations, and shows that one particularly useful capability for escape is to dive into the water. A tabulation of 9,390 prey items taken from April to August by Newton 1986 in his Scottish study area showed that regardless of the fact that the sandpiper is in the centre of the size range of prey taken, only 0.04 per cent of the sparrowhawk's prey was actually common sandpipers, so the escape tactic is effective. In fact voles and even other sparrowhawks, at 0.06 per cent, were more frequently found as prey! *Falco* species are also important. Following their pesticide-induced collapse, the Peak District was devoid of peregrines (*F. peregrinus*) until the 1980s. When the first pair returned to breed again, this was on crags overlooking the River Alport; it was watched by guardians who checked the prey, which included common sandpipers. In 1985 there were three brought to the nest: on 26 April at 07.45, 16 May at 07.30 and 30 May 11.26; clearly adult birds. Prior to that the valley generally had four or five breeding pairs, but thereafter one or two, though as this coincided with a period of shrinking range of the common sandpiper in England, this is possibly not an direct effect of peregrine predation *per se*. Nonetheless, since the main effect of predators is often for birds to avoid them, the change in a key predator in the valley will have some effect, because the sandpipers' escape tactic of diving into water is much less available on such a small stream. BNA reports that in Alaska peregrines regularly capture spotted sandpipers. The merlin (*F. columbarius*) is never seen along our study rivers, though they are up on the moors; tables of their prey items (Sale 2015) do not include *Actitis* as a component, even at the 0.1 per cent level, though in a footnote of 'other birds' one common sandpiper is mentioned.

The usefulness of diving into the water has been long known. For example:

> April 19, 1851 … I saw within thirty yards a Summer Snipe (*Tringa hypoleucos*) which was pursued by a male Kestrel, dash into the water. The hawk instead of his quarry struck the water and seemed much confused at his novel position. Disentangling himself with some difficulty from the strange element, the bird of prey flew to a tree to plume himself. When he was gone the Summer Snipe rose to the surface after an immersion of some thirty seconds, at about twelve feet from the place where he had disappeared, and flew off uttering his merry cry. (Moggridge1851)

Identification guides usually comment that a diagnostic feature of the *Actitis* is the way they fly so low that their wing tips almost skim the water.

Whether mammals are significant predators of adults is a moot point. Sandpiper response to the few stoats seen in the Peak District study is in Table 6.1: while sandpipers go crazy when they have chicks, at other times they ignore stoats. Mink destroyed eggs and fledglings on LPI, but only appear to have rarely caught an incubating adult.

PREDATION ON EGGS

A sandpiper egg at 9.5 grams for spotted and 12.5 grams for common served in a clutch of four out in the open makes a good meal. Most birds with bigger eggs are a lot more aggressive.

In the first nine years of the study on Little Pelican Island, 50 per cent of clutches hatched, but –

> 18 per cent of clutches were lost to deer mice (*Peromyscus maniculatus*); these are generalist feeders and extremely common across North America,

Table 6.1 Reaction to predators (Crows are controlled by gamekeepers such as to be hardly ever seen, and magpies are almost as scarce)

Stoat	1 July	Pair with 2 chicks, both adults alarming intensely
Stoat	29 April	Pair displaying – ignore stoat
Magpie	19 June	Magpie caught chick, adult alarming intensely
Magpie	5 June	Magpie chased off from 2 dead chicks
Magpie	12 June	Chick killed
Kestrel	22 June	Chick ringed 14 June, found by Sorby Breck group, eaten by kestrel
Kestrel	7 July	Swooped on adult and juvenile, outcome not clear
Sparrowhawk	29 April	Sandpiper dived into water to avoid hawk
Sparrowhawk	16 May	Sandpiper found at plucking post
Sparrowhawk	24 July	Juvenile found plucked
Sparrowhawk	31 May	Bird just ringed, released and as flew away taken by hawk
Sparrowhawk	24 May	Male sandpiper taken by male hawk by a narrow and shallow stream
Sparrowhawk	13 May	Sat in tree watching sandpiper
Heron	6 June	Chick ran to hide as heron flew past
Moorhen	24 June	Sandpiper with chicks alarmed at it for 10 minutes

mainly eating seeds and berries but also birds' eggs if easily found. Losses were variable but not population-threatening.

8 per cent to mink (*Mustela vison*) – all in 1975, except in 1981 one adult and clutch

0.3 per cent to otter – in 1980 one adult and clutch

6.2 per cent to birds; grackles were noted

17.4 per cent for unknown reasons.

OLM report that egg predation is greater early in season, thus impacting more on experienced birds. The means that the average rate of nest loss does not vary between age classes, because young males get later clutches when predation is less.

Prior to the start of their main study, Oring and Knudson 1972 reported 15 per cent loss to predators including garter snake *Thamnophis sirtalis*). Later an interesting interaction was found by Alberico et al 1991 when a common tern (*Sterna hirundo*) colony formed on the island in 1989. Ruddy turnstones (*Arenaria interpres*) started hanging about to predate terns' nests, and also took sandpiper eggs. In the two years 1989 and 1990, 12 early-season nests were predated by birds, and most of those were attributed to ruddy turnstones on the basis that some were known to have been taken by them, and none were known to be taken by other species; bird predation ceased when the turnstones left around 10 June for their breeding grounds in the far north. Turnstones had been stopping by in earlier years, so some had probably been taking eggs all along. The authors noted that some individual turnstones seemed more focused on eggs than others. However there was another effect of the terns, as predation by mink on sandpipers stopped because they focused instead on the terns which had bigger eggs. By and large the two predators did not overlap, as the mink did not arrive until after the turnstones had gone. Most of our losses of common sandpiper eggs are for unknown reasons, but there have been a few probably taken by magpies (*Pica pica*).

PREDATION ON CHICKS

Chicks are very good at hiding, and this is vital for their survival. After food, good hiding places are the most important feature of a breeding territory. The mortality between hatching and fledging of common sandpipers is high, and both bad weather and predation are important. The frequent disappearance of alarming adults after a day of very bad weather implies that weather may be the worse of the two. Appearance of a stoat or weasel leads to serious alarming from the adults. As well as their threat to eggs, magpies are a threat to chicks. We have kestrels, for whom a chick is as good as a mouse or vole, and one of our chicks was definitely taken in 1986. In a list of unusual prey of common buzzards (Dare 2015) a fledgling sandpiper is included. Kevin Briggs (priv. comm.), in his long-term studies of waders on the River Lune, found that little owls using nest boxes near sandpiper territories ate them. In Russia the adder has been seen taking chicks (Tomkovitch priv. com.).

Spotted sandpiper chicks have been taken by common grackles (*Quiscalus quiscula*), American crows (*Corvus brachyrhynchos*), gulls and mink. On 11 July 1981 after five years free of mink, a juvenile mink arrived on LPI on 11 July and ate most of the chicks (OLM). Although escaped mink have become a serious nuisance in parts of Britain, they were not a major issue in the locality of our study; the only year when tracks were seen and the gamekeeper killed two (24 June, 5 July) was 2005, which also happened to be a very poor year for fledging, with none fledged from the area where mink were seen.

In Yalden 1986a the stomach food analysis of four chicks was done using those caught (accidentally) by a pet dog. Breeding near humans with their dogs and cats is probably unwise, and the high density studies on islands in America do not mention cats and dogs.

As chicks get near to fledging (15–19 days old) they can also swim reasonably to avoid capture, and eventually dive and swim under water.

PREDATION ON MIGRATION AND NON-BREEDING AREAS

A well-studied peregrine site on Derby cathedral about 50 km downstream from our study site finds that after feral pigeons, a range of waders (12 species) makes up a sizeable fraction of the prey, including 51 golden plover and 7 dunlin (ten-year review in Derbyshire Bird Report 2014). No common sandpipers are reported, so maybe they fly too high or too low.

Those that I watch in winter or on migration spend their time next to the water, and when they fly they go over the water very low. I have never seen a raptor attack, and I suspect the sandpiper would just drop into the water to avoid them. I have never seen a mammal that looks likely to catch one either, though a pet dog occasionally chases one for fun. In the mangroves they may be ambushed but I have found no specific record. Throughout their range there are obviously numerous and varied predators that they have largely learned how to avoid.

Hunting is probably not serious. In BNA the comment taken from Bent 1929 is that neither hunting nor the millinery trade had any effect. They are not quarry species, they do not flock, and they are small. In the early years of ringing many recoveries in Europe were of shot birds, but this cause is now less frequent. However it has been found that subsistence catching in Myanmar was a significant factor in the decline of the tiny spoon-billed sandpipers, so there could be some local effects that are disguised in the overall large populations of sandpipers with widespread distributions.

PREDATOR AVOIDANCE

Throughout the year the main impact of predators is usually the avoidance responses that any bird needs to make to reduce its risk of death. An interesting study on western sandpipers in Canada as peregrine numbers increased showed that birds in more exposed habitat had gradually reduced their weight so that they increased their agility, and the number of sandpipers present at any time decreased as the sandpipers shortened their stay in the area (Ydenburg et al 2004).

A female sandpiper breeding in Scotland in a site with predators will leave it as soon as possible; she can move to an estuary at a time when there are very few predators and can thus afford to put on lots of weight and migrate at night to avoid them. Non-breeding-season sandpipers in mangroves seem to leave at dusk to roost communally in the open. Presumably this is to avoid night predators and disease carriers. Stanton 2013b reported high-pitched alarms, wing-raising and teetering of a common sandpiper in response to a 50 cm snake which was clearly perceived as worthy of avoidance on fish pond mud in Hong Kong.

COMPETITORS

Over the years of our study we have noted the interactions of sandpipers with most of the other birds and mammals sharing their space. Most of these are innocuous, and largely amount to 'smaller makes way for bigger'. Thus wagtails or pipits will move off from a sandpiper, and a sandpiper will give way to sheep, dippers or bigger waders. This is so on breeding, passage and wintering sites. The exception to the size rule is at the nest or with chicks, where a distraction display may be used to pull the large animal away or a vociferous attempt made to scare the competition off. Thus Nethersole-Thompson 1951 writes 'I have seen a pair attempt to drive a cock greenshank from their chicks'. To add to that, while I was by the River Lune, I saw a pair minding chicks chase a redshank away.

A bizarre and contrary observation is a willow warbler attacking a common sandpiper in Zimbabwe (Vernon 1965).

On wintering areas in mangroves Zwarts 1985 found them sharing the mud with crabs and whimbrel. The small crabs were food for the sandpipers while a big crab could hurt a sandpiper but be food for a whimbrel. Where I watch them in winter on tidal rivers in the south of England, there are very few other birds on their shore; an occasional little egret (*Egretta garzetta*), clearly eating different food, or a wagtail. My suspicion is that the sandpipers do not feed further downstream because they would be in competition with redshanks. On passage I have watched a sandpiper steadily feeding along an edge till it gets close to a redshank and then take avoiding action.

PEOPLE

One serious animal competitor is the human (and associated pets). Sandpipers are adaptable to a limited disturbance; as adult birds they can fly elsewhere, and on the nest they can sit very tight and be invisible, while with chicks they can hide and creep away unseen. By an isolated cottage they may find the garden a useful feeding ground. But continuous disturbance, and the presence of cats, dogs and rats mean that towns and suburban sprawl and high-usage recreational sites are not used.

As an individual each sandpiper makes a response to the presence of people. The Peak District National Park has high numbers of people (tens of thousands live and work there, and there are about 20 million day-visitors per year) and the effect of their disturbance on wildlife is important. Yalden & Yalden 1990 did a detailed study of the effects of people on the Pennine Way, a busy long-distance hiking route passing through golden plover habitat; the study contributed to the decision to create paved sections through the most fragile habitat by air-lifting in natural stone slabs, which kept people from wandering off, damaging the vegetation and disturbing birds. The birds got used to people on the path. Yalden also studied the disturbance effects on common sandpipers around the reservoirs. Two main categories are fishermen and casuals (e.g. having a picnic): the former stand fairly still for long periods, and in many sandpiper territories outnumbered the latter. The results were fairly clear (Yalden 1992a). On the side of the reservoir where there was easy access due to the road there were more people and fewer sandpiper territories. The sandpipers per kilometre seen on a visit were inversely correlated with the number of people per kilometre. Also the presence of the busy road restricted the options for mobile chicks to retreat far from the reservoir edge. On the opposite side of the reservoir there were fewer people and more sandpipers. There, public access was a rough track with hardly any traffic, and if there was a busy time (e.g. a fishing match) the sandpipers could retreat into woodland (and as we have seen in Chapter 5, they find wood ants good to eat). If a fisherman happened to put his paraphernalia onto a nest with eggs, that would be unfortunate, but we have no knowledge of this ever happening.

These observations done as a specific study were consistent with some casual observations made in 1979–82 (Yalden 1984) when a change to access was made up the Derwent Valley in early 1981. Where a car park was closed a sandpiper took over, while four territories that more people went to as a result of the changes became deserted by the birds.

Watching through a telescope from afar, Yalden was able to see that as a person walked by a chick it hid for 3.1 minutes on average, and in one case a succession of walkers led to one hiding (and thus not feeding or being brooded) for 34 minutes. An adult sandpiper starts alarming when a person is typically 75 metres away and then flies at 27 metres, often into a neighbour's territory so a bout of display ensues thus wasting energy (of both families) which needs to be made up by feeding rather than guarding. Observations of feeding showed that the adults near the path spent 8 per cent more time feeding than those observed on the streams which are free of disturbance. Young chicks are brooded roughly every 8 minutes so all these interruptions can add to a significant effect such that a territory with a pathway through it will eventually become untenable when the path is used by lots of people. Even if only a few people are present, the adult sandpiper going into another territory will provoke a response, and fights of 15–27 minutes were seen as a consequence.

These observations from afar obviously lead to the question of how much effect our own activities watching and ringing the birds has had. It will have had some. Our aim is to catch a bird only once and mark it with colour rings; then it can be observed from afar for the rest of its life. They rarely get retrapped, as they now know what we are up to. Probably some new individuals who are considering breeding but then get caught decide to leave the area and go on to somewhere else. The regular breeders can probably distinguish us who follow them from the farmer who ignores them. Overall we hope the effect of our activities is very small, otherwise we have wasted 40 years.

BNA reports that blood sampling caused 1–2 per cent harm, and leg bands caused 2.6 per cent foot loss. In common sandpiper studies we find foot loss (mainly from sheep's wool entangling around toes) on un-ringed birds as well as ringed birds (again around 2 per cent). Since the Americans were so efficient at catching all birds soon after arrival it may be that around that 2.6 per cent would also have occurred in an un-banded population.

In trying to assess the overall effect of human recreational disturbance it was concluded that the success rate of those birds that did choose to breed on the disturbed bank was no different from those on the opposite bank so their response behaviours of being at a lower density and taking avoiding actions worked. They are intelligent enough to assess the situation, and anglers are already active in April when the birds arrive. Occasionally we made a polite request asking people not to picnic by a nest and in the 1980s we produced an information leaflet for anglers, to make them sensitive to the needs of ground-nesting birds. Box 4 in Chapter 1 shows the stress that a spotted sandpiper had with too many people.

The classic reported case of excessive recreational pressure on breeding waders is around Loch Morlich in Scotland, where up to 1945 there were eight wader species which bred; a total of over 40 pairs of which at least 20 were common sandpipers. By 1985 the only shorebirds left were two pairs of common sandpipers, because car parking took habitat, continual recreational disturbance kept birds away and the vegetation was trampled (Watson et al 1988).

On migration and the non-breeding grounds, sandpipers will typically fly off at about 30 metres but people going about their normal activities are not a massive problem for them. A systematic study in Ghana (Gbogbo et al 2013) showed this: it seems to be one of the most tolerant waders, as in the second largest area of mangroves in India a major decline of waterbirds over 20 years was reported due to increasing human disturbance,

Figure 40: Up a mangrove channel, sandpipers ignore the tourist boat and the local people harvesting.

cattle grazing, fishing, bird hunting and recreation – but common sandpipers seemed unaffected (Sandilyan et al 2010).

OTHER NUISANCES

Pesticides – BNA reports that PCBs, DDE and dieldrin were found in spotted sandpipers, but there is no evidence of any effect on breeding or mortality.

Parasites and disease – On the breeding grounds, all birds caught appear to be very healthy. Derek Yalden found one chick in 1990 that had two fleas (male *Ceratophyllus garei*), and in 1995 an adult with two louseflies (female *Ornithomya chloropus*), but not apparently life-threatening. Finding an individual that has died of disease is so rare that the find of an upland sandpiper (*Bartramia longicauda*) on the breeding grounds in Kansas warranted a whole scientific paper describing it (Sandercock et al 2008). Its post-mortem deduced that it had died of a reovirus infection (mainly known from domesticated poultry and parrots). It was a radio-tagged male bird that had been looking after its chicks on 16 June; apparently fit, but found dead two days later. The other dead radio-tagged birds (seven of them) found over the six years of the study had all been predated.

Piersma & van Gils 2011 show that investment in the immune system is likely to be less in a shorebird that breeds in the Arctic and winters in marine environments, as both are low in pathogens, while ones that spend their time in temperate and tropical freshwater will need to invest more in their immune system. Intestinal Helminths (worms) have been widely studied in spotted sandpipers throughout their range; 36 different species are listed in Didyk et al 2007, but whether they do much harm is unclear. Shorebirds are reported to be remarkably low in blood parasites (haemosporidians), with less than 1 per cent of a sample of 318 of various species having any at all (Soares et al 2016). As in most other aspects of their lives *Actitis* appear to be average, and there is no evidence known to me that disease is of particular importance relative to bad weather, predation and old age. But if they were dying of some disease, especially in the tropics, it would be difficult to know.

Nonetheless, in earlier chapters we have seen birds on migration that do not put on weight like their companions, we have seen birds not moulting fully, and ones not making the journey back north to breed, so it seems plausible that sickness is involved.

Kleptoparasites – We have not seen other creatures stealing food, nor found any reference to such behaviour in the literature.

7 POPULATIONS

WORLD POPULATIONS

Both common and spotted sandpipers are widespread and abundant. The end-of-year common sandpiper population is judged to have 1.5–2 million individuals from across north, west and central Europe, more than 1 million in eastern Europe and western Asia (Delany et al 2009), and I guess about another 1 million further to the east. In Chapter 3 it is estimated that over 3 million can be accounted for across their wintering grounds.

The spotted sandpiper population was estimated as 150,000 until a re-evaluation in 2012 (Andres et al 2012), when it was assessed as 660,000.

ASSESSING BREEDING POPULATIONS

When we estimated the population of the British Isles as of 1990 as 30,600 pairs (Dougall et al 2004) this was based on several very detailed county avifaunas and numerical surveys of many different kinds in countries where we have a ratio of around 20 birdwatchers per breeding sandpiper. But we pointed out that extreme estimates of 11,500 or 51,500 could also be argued. One of the difficulties is that where there are lots of people there are few sandpipers, and in the areas where there are lots of sandpipers there are far more iconic birds to watch during the very short season that sandpipers are present. Thus breeding estimates across the vast and low- peopled areas of Canada and Russia will not be very precise.

Even where there are detailed observations

- Counts in 'suitable' areas give widely different densities. Thus NT2 tabulated 18 reported densities ranging from 0.1 to 6 per km (and an outlier at 25 on a very short stretch).
- In an area that remains nominally unchanged, the number of pairs may change considerably over a few years.

In the European Bird Atlas (Hagemeijer & Blair 1997) Finland had a population of 240,000 breeding pairs. If each pair still had a chick alive at the end of the calendar year, then Finland alone would contribute 720,000 to the world population of common sandpipers – a greater number than the whole world estimate for spotted sandpipers. The areas of Ontario, Minnesota and similar lake-rich areas of North America appear on the map of the world to be about 10 times that of Finland, and populations are reported down the Rockies, so that raises another question, as to whether even the recently increased estimate for spotted sandpipers is too low or the Finland estimate too high. Clearly the likely alternative is that *Actitis* are more than 10 times less densely distributed in American habitat than in the European.

One good thing is that they are not cryptic. Three visits during the breeding season are sufficient to estimate breeding numbers defined as territorial pairs. This was based on work on streams in the Pyrenees between 1998 and 2000, and showed that just one visit detected 83 per cent of territories (D'Amico 2002). This is consistent with our general findings that they are easy to census when breeding (Yalden & Holland 1993). Populations of breeding pairs for various countries have been given as Finland 240,000, Norway 140,000, Sweden 110,000, Estonia, UK and Belarus 20,000 each, Spain, Latvia, Slovakia and Romania 10,000 each, and another 22 countries adding about 20,000 more (Hagemeijer & Blair 1997). To get the end-of-year populations the number of pairs is usually multiplied by three to allow each pair to have one surviving offspring, thus giving 1.8 million at the end of the year.

GENERAL TRENDS IN POPULATIONS

On the world scale, both species are of least concern on the Birdlife International scale.

In the British Isles the decline in the breeding season has been such that it is on the amber list (along with 95 other species and 67 that are on the red list; far more birds are red plus amber than green at 81; Birds of Conservation Concern 2016). Two trends are clear. Firstly in British Isles Atlases, the range shrunk from 1968/71 to 2007/11, with the biggest effect in Ireland. Secondly, annual counts show a noticeable reduction in the numbers counted from 1985 to 2015. These trends mainly reflect attrition round the edge of its range, and counts do not include lochs and reservoirs in the core areas. Losses around the edge are reported in other countries on the edge of the range (e.g. Switzerland).

The numbers of common sandpipers across Scandinavia showed no change between 2006 and 2013 (Lindstrom et al 2015), though there had been a decrease in Sweden beforehand. Numbers on migration do not appear to be generally changing at big sites with long series of data (see Chapter 2 above). The Pan-European Common Bird Monitoring System (PECMBS) shows a downward trend of around 35 per cent from the 1980s to the present.

Spotted sandpipers have been reported lost from some areas, but in core areas the change is less. In Quebec no evidence of change was reported over 100 years (Gauthier & Aubry 1996), whereas the North American BBS reported an overall decrease of 1.35 per cent per year from 1966 to 2015 (Sauer et al 2017).

HOW DO SANDPIPER POPULATIONS WORK?

The point of colour ringing a bird is that you hope to see it again. Indeed you hope to see it many times, and as well as the sort of behaviours that you can now link to sex and age and its previous history, the sum of life histories shows how a population sometimes increases and sometimes decreases, but hopefully sustains itself. As well as the faithful birds which are seen multiple times, often year after year, there are others that are more erratic and over a short study are essentially a nuisance; but over a longer period the patterns give further insight into the population dynamics. The simple categories are

'Never seen again' – it is caught in a breeding territory in the breeding season, but then is lost to science

'Lazarus bird' – it is seen (maybe a few times) in the year of ringing then does not return for some years but later 'returns from the dead'

'Known to move' – the individual is found, sometimes years later, breeding elsewhere.

The never seen again birds tend to be in early and late season, and the fact that some of these are on migration was confirmed in 2011 when one was photographed on 5 May, then 150 km further north on the Tees on 6 May. A good example of a Lazarus bird exploring and later settling was a male caught on 20 April 1982; he was on our 'never seen again' list until he appeared again on 2 July 1985, and then from 1986 to 1988 he bred successfully in the study area.

Those adults that have moved are described in some detail later. There is a continuum of movement from just moving to the next territory, to moving over a kilometre though still within the study area, to moving to another place in the Peak District, to moving a long way.

In the Ashop study area (described in Appendix 1) the number of pairs has varied over the 40 years, as seen in Figure 41. Previous to starting colour ringing in 1977 the population was perceived as high and fairly stable, and results from casual ringing showed that the birds were fairly long-lived and site faithful; it was those observations that led to the decision to start the project. The challenge has been to find the causes of the variations that have happened from year to year.

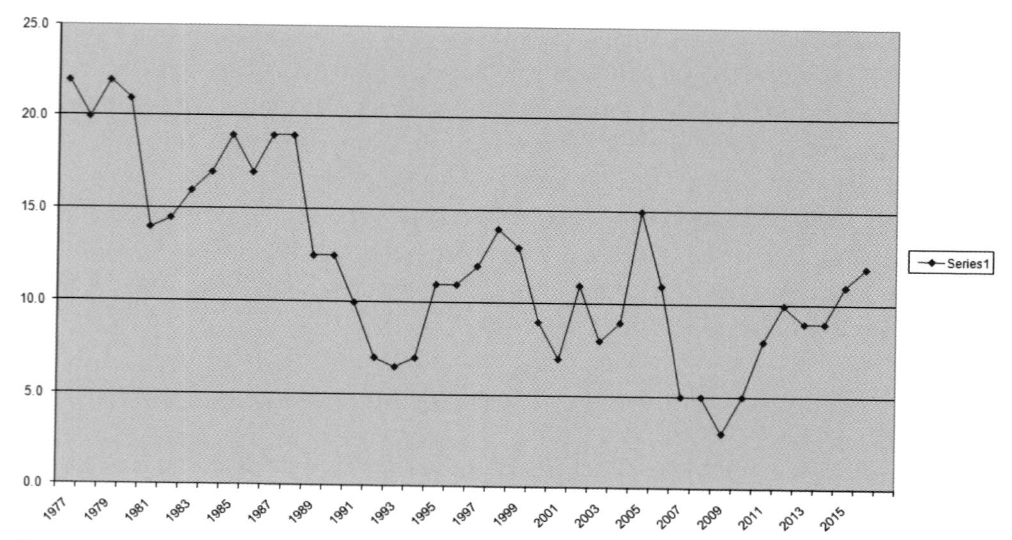

Figure 41: The variation in the number of occupied territories in the Ashop study.

For the world population the key things are adult survival overall and their success in producing young that join the breeding population wherever they do so.

RETURN RATE AND ITS RELATION TO ADULT SURVIVAL

In our best year from 1979 to 1980, the return rate of territory males to the Ashop was 0.94. In our first paper (Holland et al 1982a) we quoted a 'survival' of 0.81 based on the return rate during our first few years of study. In Holland & Yalden 1991b we had followed all the original colour-marked birds to the end of their time in our study area. The period included one year of very poor return (attributed to appalling local weather); we reviewed various ways of analysing the data and gave a 'survival' of 0.79. During the 1990s (Holland & Yalden 2002) several birds deserted the Ashop for reservoirs, but the 'survival' was 0.84 using Haldane's method allowing for desertion. In Holland & Yalden 2012 we compared the return rate of territory holders during the period of high population from 1978 to 1987 (at 0.79) with the average return rate during the ten-year period of low population from 1998 to 2007 (at 0.63). The BTO runs an excellent scheme called Ringing Adults for Survival (RAS) to which our and Tom Dougall's data from the Scottish study area (Appendix 1) were sent, and their computer says that adult survival is 0.59 for males and 0.61 for females.

So from this range of numbers (0.94 to 0.59), which should we choose as a relevant adult survival for the world population?

Clearly, if we tried to measure survival by only looking at the return rate to one territory, then as soon as a bird moved even the very short distance to the next territory it would appear to have failed to survive. Thus the size of population that is being studied is important, and whether it is likely to be 'self-contained'. The Ashop population was around 20 pairs through the 1970s and 1980s, and the upper parts of the streams are separated from the next catchment over the hill by habitat that was good for grouse and moorland birds but not for sandpipers. The most downstream territories are separated by a stretch of low-suitability river before the reservoir. From the time when we started ringing in 1968 to the start of our detailed study in 1977, several retrapped individuals had shown good site fidelity and long life. The return rate of 0.94 from 1979 to 1980 led to a mindset that they were totally site faithful. There was extraordinary weather in 1981, when very heavy snow in late April completely blocked our access; there was report of a lapwing found as the snow cleared sat on its eggs frozen to death, and our only ever local finding of a dead sandpiper. The population of sandpipers dropped and the return rate was low, leading to a mindset that these had all died.

However there were more and more observations over the next decades that there was regular desertion of the area by adults. The main known destination was the adjacent reservoir complex, but this is not the only destination as other reservoirs are much less well watched. Figure 42 shows some long movements of adults. Why an area might become unattractive and lead to desertion is discussed later. But as many adults clearly deserted, the drop of return rate from 0.79 to 0.63 over 20 years suggests that the real survival rate is greater than 0.79. Taking another approach, the return rate of similar-size migratory shorebirds like dunlin, sanderling and purple sandpipers to highly stable wintering habitat

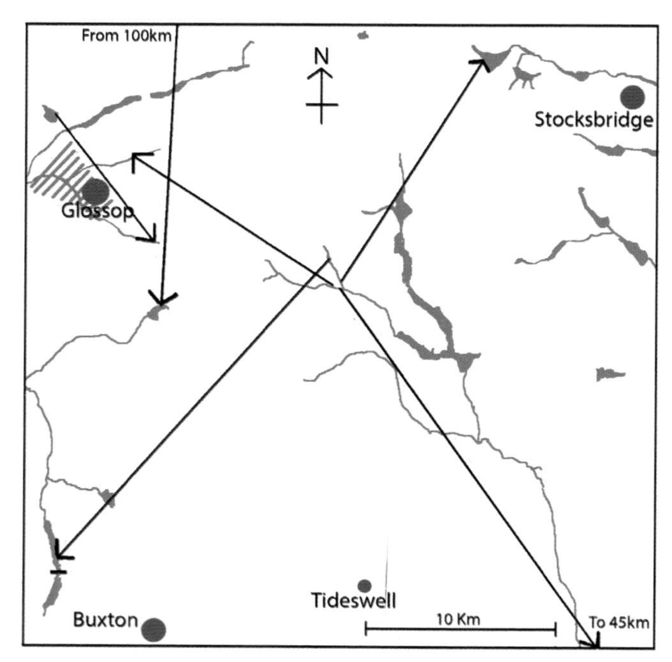

Figure 42: Known movements of adults from catchment to catchment and out from and into the Peak District (these are in addition to those moving between the Ashop and the LDH reservoirs described later in the text).

are reported as around 0.8. So I think that a reasonable value of adult survival is over 0.79 for common sandpipers, and will use 0.81 as we first derived. This is essentially the return rate of territory-holding birds coming back to an attractive assemblage.

For spotted sandpipers similarly, Reed & Oring 1993 suggested that a good estimate of survival was the return rate of experienced and successful adults, which over the 18 years of their study was 0.63. The fact that this is less than for common sandpiper is reasonable, because it is a smaller bird. They elsewhere suggest that the multiple laying resulting from polyandry is very stressful for the females, so that factor too may shorten life. The oldest spotted sandpiper is reported as 12 years and common sandpiper as 15 years, which is broadly consistent with the different estimates of survival. (Large totals of each species have been ringed/banded.)

PRODUCTIVITY PER FEMALE

We have seen in Chapter 1, in our Ashop study (Figure 13) the fledgling productivity per territory was around 0.8. We thought that this was unduly low, and may have been the cause of the Ashop decline. Thus it was useful that Tom Dougall started another detailed study in Scotland, away from the retreating edge of the English range. We looked at whether productivity was better in the Scottish area, but found it was much the same (Dougall et al 2005). So maybe it is higher on lochs and lakes. Derek Yalden had started to do annual surveys around the reservoirs as part of his long-term monitoring of all sorts of creatures in the Peak District (grouse, plovers, geese, deer, wallabies, hares, heather, bumblebees, people, dogs). From these surveys it is also possible to get a very rough estimate of fledglings of around one per year per pair; not, then, statistically differentiable

BOX 5

BRUCE FROM BIRTH TO DEATH

It is quite unusual to know not only the place of birth, but also some breeding history and the place of death on migration, but Tom Dougall described one such bird in *RAS News* in April 2005. This male, which I will call Bruce, was one of a brood of three that had been ringed on 8 June 1996. Bruce was next seen on 9 June 2002 breeding 6 km away from his birthplace, mated to a female that had been ringed 3 km away in 1999 as an adult. They bred together in both 2002 and 2003, hatching three chicks in both years. But then on 22 April 2004 Bruce was found dead in the south of England on northwards migration (Chapter 4) about 600 km short of his destination. His previous mate found a new male, and when caught on 8 May was gravid, but her nesting attempt failed.

If Bruce had been raised 6 km in the opposite direction outside the study area and his 2002 female also come from 3 km outside the study area, they would have been new to the territory – so it would then have been assumed that they were both one year old when in fact they were six years and at least four years old. He would have been put down as only breeding for two years. Movements of females between breeding attempts are even more common, so the use of the return rate as a surrogate for individual survival is unreliable. This female may well have moved to a new territory outside the study area following the failure of her mate to return, and by the time she died could also have been around ten years old.

from our 0.8. As we have seen in Chapters 2 and 3, there are plenty of juveniles which reach the non-breeding areas. The ratios of first-years to adults caught would imply that the annual productivity across the core areas of Europe is higher than one per pair, so in world terms using one is a reasonable number (possibly it is higher)

For spotted sandpiper on LPI the female has more than one male working on her behalf; experienced females produced 2.32 fledglings per year and inexperienced 1.48, with an overall running average of around two per year.

FIRST-YEAR SURVIVAL FROM FLEDGING IN JULY TO JOINING A BREEDING POPULATION IN MAY

In shorebirds that are well studied in non-breeding areas of Europe the over-winter survival is generally high unless there is extreme and long-lasting cold weather. It is thus reasonable to assume that the mortality is low in the warmer tropics, where they stay for as long as they reasonably can. But the young bird will probably have been excluded from the best habitat so will have a harder life, and it is not going to be more successful on northwards migration than an experienced adult. Furthermore, once it gets back to the breeding area it has to find a territory it will try to occupy. All the indications are that this period in April is a crucial test of an individual's survivability.

Of the 169 juveniles ringed with a full colour code that have left our Ashop and LDH areas, 25 per cent are known to have survived because they have been seen in later years somewhere; the figure for the spotted sandpiper was 17 per cent (Reed & Oring 1993). It is not unreasonable to assume that as many again remain undetected and thus to guess a very rough first-year survival of around double those known to return (so, 50 per cent for common and 34 per cent for spotted). Another approach that could be taken is to arbitrarily assume that first-year mortality is twice that of an adult, and that would give survivals of 62 per cent and 26 per cent, but this rate is unreasonable for spotted sandpipers because it is so small. For common sandpipers, the survival rate that we guessed in our first attempt was 57 per cent (Holland et al 1982a), and we have continued to use 0.57 as we have not had any compelling reason to change it. Given the slightly smaller size of the spotted sandpiper, a consistent guess would be 0.5.

It must be emphasised that in old literature there is sometimes the statement that they do not breed at one year old, but all detailed ringing studies have shown that they do. Thus of the 14 locally bred spotted females that entered the LPI population, 10 did so in their first year and the other four probably did breed, but elsewhere. For 13 males the numbers were 4 in first year, 4 in second, 2 each in third and fourth and 1 in the sixth, but again their mobility would support the idea that they were breeding somewhere else in their first year. Holland et al 1982a reported on first-year breeders, and this has remained a regular occurrence in all studies of breeding common sandpipers.

AN OVERALL BALANCE {APPROXIMATELY}

If we start with 100 adult common sandpipers in May one year and they have a survival of 0.81, there will be 81 still alive a year later. They will have produced around 50 fledglings (1 per female);if the juveniles survive July to May with a survival of 0.57 (making 28 first-years), that will make up for the loss of adults (19 lost: 28 gained). The margin would imply that there is scope for population growth – but quite small changes to adult survival, fledgling production (e.g. allowing that first-year birds are less productive) and first-year survival soon reduce the margin to zero.

If we start with 100 adult spotted sandpipers in May there will be 63 a year later. They will have produced 100 fledglings (two per female) and if those fledglings' survival is 0.5, that will make up for the loss of adults (37 lost: 50 gained). In Chapter 1 we saw that if a female only has one mate she will produce 1.22 fledglings, so experienced, good females with good territories producing over two fledglings per year are important to keep the population going.

FLUCTUATIONS IN A LOCAL POPULATION FROM YEAR TO YEAR

Figure 41 showed the observed population from year to year in the Ashop study area. Over many years we looked at correlations with this and that, to try to see what might be driving the variations. One obvious one is whether a good local production of fledglings led to an increase the next year, but it did not. The temperature in late April looked promising for a few years (Holland & Yalden 1991a), but did not withstand the testing over a longer period with a normal range of weather (though it is undoubtedly true that an extreme event, such as the long spell of snow in late April 1981, can depress the local population, as would other freak events). An attempt to correlate with the winter North Atlantic Oscillation showed a slight (but statistically non-significant) effect, where a high NAO (warm, wet and stormy in north-west Europe and cool and dry in north-west Africa) had a lower return rate (Dougall et al 2010, Pearce-Higgins et al 2009). This was consistent with the idea that getting from West Africa to Europe at the end of winter is a critical time.

Recruits may come from:

1 local fledglings
2 fledglings from elsewhere
3 adults moving in (thus deserters from elsewhere).

We know that the first is a small proportion, because we ring most of the young and the numbers that have returned to our streams is small. One of the key interests in comparing Tom Dougall's Scottish study and the Ashop study was to see if the low number of local first-years was why the Ashop population was going down, and indeed the number of chicks returning to breed in the Scottish study area was higher there (Dougall et al 2005). But also the detailed geography is different; the streams do not feed a reservoir complex that provides attractive alternative sites, and the overall 'target area' in the Scottish study area is bigger than the Ashop. A reasonable compromise interpretation is that in any breeding population of 25 pairs which will need 10 new recruits per year, 2–3 of those will be sourced from that population's previous year production.

For sources 2 and 3, a problem is that it has not been possible to be sure whether a new bird is in its first breeding season. There are some plumage indications that a bird is

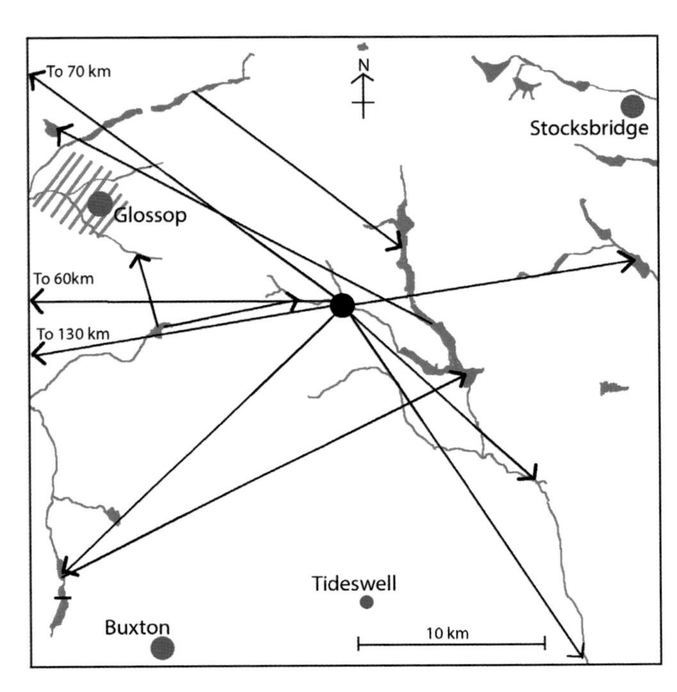

Figure 43: Known movements of chicks from their Peak District birthplace to a breeding place. Longish movements are also seen in BTO recoveries e.g. a Scottish chick from Loch Nevis in 1934 was found dead at Loch Ewe 60 km away on 3 June 1935, and a Welsh chick from Rhayader in 1970 was found dead near Harlech 75 km away on 16 May 1971, so the movements mapped here are not peculiar to the Peak District. Movements between the Ashop and the LDH reservoirs are not mapped.

first-year if it has not completed its moult (Meissner et al 2015 and Chapter 3), but these have only been developed during the latter years and all of our new recruits have just been aged as adults. We know that fledglings from other areas move in as well as ours going elsewhere (see Figure 43).

We know that we have birds ringed as adults one year that have moved to a new breeding place in a later year (Figure 42) and some of these have bred for several years in the same place before deciding to leave. They may move to and fro – see Box 6. Thus it is clear that there is not unquestioning site fidelity. Similarly when Reed & Oring 1993 looked at USA recoveries of spotted sandpipers, three adults that had been ringed in breeding areas were found breeding in later years 61, 113 and 147 km away.

Thus each individual makes a decision each year on its breeding place, and makes it quickly because it only arrives at the end of April and so has just a short time to breed. An idea explored in Holland and Yalden 2012 was that all birds are attracted not just by habitat but by the presence of other birds. If a bird comes back to its territory of last year and within a day or so its previous colleagues have not returned, it can move on. On the other hand any new bird coming to an area that (as well as being good habitat) is alive with sandpipers will be attracted to that area. Another part of this process is the reconnoitring that a bird does at the end of the season, checking out neighbouring areas to see how successful they have been in producing fledglings. Oring & Reed 1996 came to the same conclusion about the importance of reconnaissance in the spotted sandpiper; their statistics suggested that a transient female one year had a higher chance of joining the population the next year if on reconnaissance there was a high population of males looking after eggs and chicks (that is, it was a thriving community).

Support for these sorts of behaviour came from birds ringed but 'never seen again' in the Ashop. Below are the numbers of male and female common sandpipers caught and colour-ringed in the Ashop study area but never seen again, according to catching period:

	24 Apr–7 May	**8 May–30 Jun**	**1–10 Jul**
Males	5 (0.4 per day)	8 (0.15 per day)	6 (0.6 per day)
Females	2 (0.15 per day)	17 (0.32 per day)	0

At the start and end of the season, males are looking for good places to establish a territory, while by the second week of May females are looking for good places with settled males; by July the females have departed. Some males that have been territory holders from a previous season have been seen early on in the season, but then they disappeared (16 individuals between 20 April and 7 May).

There is an apparent tendency for new birds to take up territories next to experienced breeders. In Holland & Yalden 2012 we gave the table, below, here Table 7.1, on the choices made by new male common sandpipers in taking up available territories in the Ashop study site in relation to the status of adjacent territories and their occupiers.

BOX 6

A LONG-LIVED FEMALE COMMON SANDPIPER BEING FLEXIBLE OVER 13 YEARS

BV45510 was an adult female ringed at the Ashop football field (territory 8) in 1980; she had chicks, and she was last seen on 26 June as they neared fledging. The next year she moved upstream to the next but one territory, no 10 (by the farm) and again hatched chicks and returned successfully to the farm for two more years.

In 1984 she was not seen until June, when she was found on a reservoir territory.

In 1985 she was back at the farm and appeared to be successful each year to 1987; then in 1988 she moved back to downstream from where she started into territory 7.

In 1989 and 1990 she was back at the reservoir, appearing to be successful.

At the start of 1991 she was seen at reservoir, then in mid-May at the bottom territory (1) in the Ashop on one day only, but where she bred that year is a mystery.

In 1992 she appeared again, successful, alarming with chicks at the reservoir on 20 June.

A LONG-LIVED (15-YEAR) MALE COMMON SANDPIPER

NV54164 was ringed as a chick about a week old at Ladybower Reservoir on 21 June 1992. He occupied a territory at the top of the Ashop in 1993. In 1994 he moved down to Ladybower reservoir to a territory on the opposite side to where he was born, which he occupied for 14 years, being last seen on 19 June 2007.

The breeding success of reservoir birds was not systematically checked, so the lifetime reproductive success of these two birds is not known. The next longest-lived male, which was at territory 8 from 1977 to 1988 (12 summers), raised 19 to fledging.

Table 7.1 These data do not meet any statistical test, but it was a way of indicating that being surrounded by birds, particularly those that are experienced, seemed to have some attraction for a new bird; occasionally a pair of new birds would even force themselves into a tiny territory between two occupied territories. Success from hatching to fledging was looked at in Holland & Yalden 1994: pairs with no neighbours failed to get any of their chicks fledged in 41 per cent of territories, while those with neighbours on both sides only failed in 19 per cent, though the statistics were also too few for significance.

<table>
<tr><th>Status of adjacent territories</th><th>Status of neighbours (N = new, R = returning)</th><th>Territories taken up / No. available (proportion)</th><th>Territories taken up / No. available (proportion)</th></tr>
<tr><td>Both vacant</td><td></td><td></td><td>27/131 (0.21)</td></tr>
<tr><td rowspan="3">1 vacant, 1 occupied</td><td>2 N</td><td>10/38 (0.26)</td><td rowspan="3">43/172 (0.25)</td></tr>
<tr><td>1 R + 1 N</td><td>13/57 (0.23)</td></tr>
<tr><td>2 R</td><td>20/77 (0.26)</td></tr>
<tr><td rowspan="5">Both occupied</td><td>4 N</td><td>1/2 (0.5)</td><td rowspan="5">41/83 (0.49)</td></tr>
<tr><td>3 N, 1 R</td><td>3/10 (0.3)</td></tr>
<tr><td>2 N, 2 R</td><td>8/18 (0.44)</td></tr>
<tr><td>1 N, 3 R</td><td>18/30 (0.60)</td></tr>
<tr><td>4 R</td><td>11/23 (0.48)</td></tr>
</table>

At the other extreme there is an expansionary aspect in their searching. There is the major part of England, generally deemed unsuitable for common sandpipers, yet in this unsuitable part of the country almost every county had a breeding record in some localised place in some year. So something (e.g. the cessation of gravel extraction leaving bare beaches) made it suitable enough for a pair to colonise, and these observations are given in Chapter 9 where we look at habitat. Many of these sites are over 100 km from the nearest regular populations, showing that there is a strong searching and exploratory urge for trying new places. Just taking one of those places – that near Horsham in Sussex in 1978 was 250 km south of the normal range, yet this pair had chicks on 23 May which is very early, and suggests that they were experienced adults; another possibility is that they had wintered fairly locally (or they could have been both). The Arun that flows by Horsham is one of the rivers which have a few sandpipers wintering near the coast (about 40 km from the breeding site). But these isolated sites rarely get used for more than one year; it appears to be a totally random occurrence that male and female arrive at the right time for each other.

So, we see a decision-making process for an individual who, at the end of a long intercontinental migration, aims for the place it had come from. As a returning adult it has a potential breeding territory to return to, but also knows that there are other places it has seen during previous summers. If it gets back and the habitat is still good, and the weather is normal, and there are other birds to mate up with, then site fidelity usually occurs. If the habitat and/or conditions are too bad it moves on. As a first-year bird it recognises appropriate habitat, and then has that perception reinforced by seeing that it is occupied by excited birds of its own kind. It has usually travelled alone and needs to find a mate.

But the number of recruits will depend on last year's breeding success over quite a large area, perhaps 50 km in radius. It is also possible that places on the southern edge of the range are reinforced by more successful populations further north.

CAN WE EXPLAIN ANY OF THE VARIATION IN THE ASHOP POPULATION?

Potentially there is a plausible story, but we need to look at a few other clues first.

STREAMS ARE PROBABLY A SECOND-RATE HABITAT FOR BREEDING

After 40 years of monitoring the Ashop, I have also come to think that a better place to study common sandpipers would be a good loch or a good reservoir. In the next subsection we will look at the interaction between the Ashop population and that of the reservoirs it feeds. But before that we look at some other data on streams.

Lune tributaries

Sedbergh School did several surveys of part of the River Lune (shown in Figure 48 in the next chapter) and various tributaries upstream from Kirkby Lonsdale up into the fells totalling 96 km (Cuthbertson et al 1952). Ingram Cleasby, who had been at Sedbergh School doing the surveys as a boy, renewed his interest on retirement (Cleasby 1994) and commented that 'the absence from small streams in the 90s was painfully obvious'. He also noted that even on good stretches of stream there appeared to be large variations, and a stretch of the Rawthey that had sixteen in 1951 had only had nine in 1939 and was down to eight in 1991. Kevin Briggs (oystercatcher enthusiast 1970s to the present day; priv.comm.) and I did a range of Lune surveys in various years up to the present; these give a picture of a variable and slightly declining population on the whole river, and a bigger reduction on small stream tributaries. Kevin's results for the Lune upstream from Kirkby Lonsdale 1978–79 were 74 territories, but for 2007–09 down to 40 territories. The Barbon tributary had 9 territories in 1951 but I only found 2 in 1990, with the road up the valley now being a picnic zone.

Small streams in the Peak District

When we started our project we reported on the whole population of the Peak District (Holland et al 1982b) which we reckoned as around 200 pairs. Most were around reservoirs, but 78 were on streams and rivers. We have not done a systematic repeat survey, but ad hoc observations would suggest a population now approximately halved on streams.

Cheeseden Brook (a stream ultimately feeding the Mersey)

A detailed local report of streams around Heywood (Whittaker 1932) claimed six pairs along Cheeseden Brook; by 1980 there were none.

Generally many areas from which declines are reported in all sorts of county bird books seem to be on streams. However let us do a thought experiment on Figure 41. Observer A visits the Ashop in 1981, 1990, 1998, 2005 and 2016, recording 14, 12, 14, 15, 12, and telling their friend, Observer B, that the population has been steady. Observer B goes in 1979, 1990 (with A), 2001 and 2009, and tells A that they're crazy as the population has crashed from 22 to 12 to 7 to 3! So streams are very variable.

THE LADYBOWER-DERWENT-HOWDEN {LDH} RESERVOIR COMPLEX IS ATTRACTIVE

Soon after we started the Ashop study, Derek Yalden started doing a full survey, on about 20 June, around the reservoirs into one of which the Ashop runs (see Appendix 1). In some years he also did it in May and in one year he carried on through the season so that he could be confident that the June count was reasonably representative.

Figure 44 shows the variability of that population superimposed on the Ashop population. There is no long-term correlation. By eye, one might say that both had a period of increase in the mid-1980s followed by a drop; that something attracted extra birds to the reservoirs after 2005 and whatever that was may have sucked birds from the Ashop to give its low point in 2009. The attractive force may have overshot the normal capacity of LDH and led to the bounce-back of the Ashop population from 2010 through to 2016. A chick from Ladybower in 2012 that was in territory 17 (on the narrow Alport tributary) in 2014 stayed to 2016 (at least); on previous experience it would not have been expected to stay on such a marginal stream, so its presence was a good sign for population increase though not evidence of any great weight.

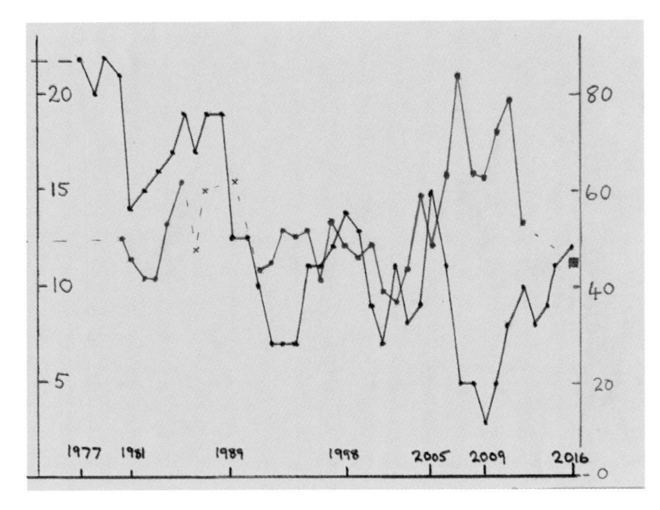

Figure 44: The variation in the Ashop population (blue) and the LDH reservoirs (red). The LDH data is from Derek Yalden's surveys except the 2016 point which I did, and possibly is low as some difficult-to-access areas I dodged.

But overall –

a) The population of LDH is less variable (max/min = 2) than the streams (max/min = 7), and the valley with reservoirs has more pairs than the valley with streams.
b) The movement of adults has been from the streams to the reservoirs (25 to the reservoirs, versus 3 from them).
c) Some chicks from LDH have bred up the streams in their first year then returned as adults downstream (e.g. Box 6 male). The net movement is that 9 have gone back down (3 to the River Derwent below the Ladybower dam) and 4 have stayed up.

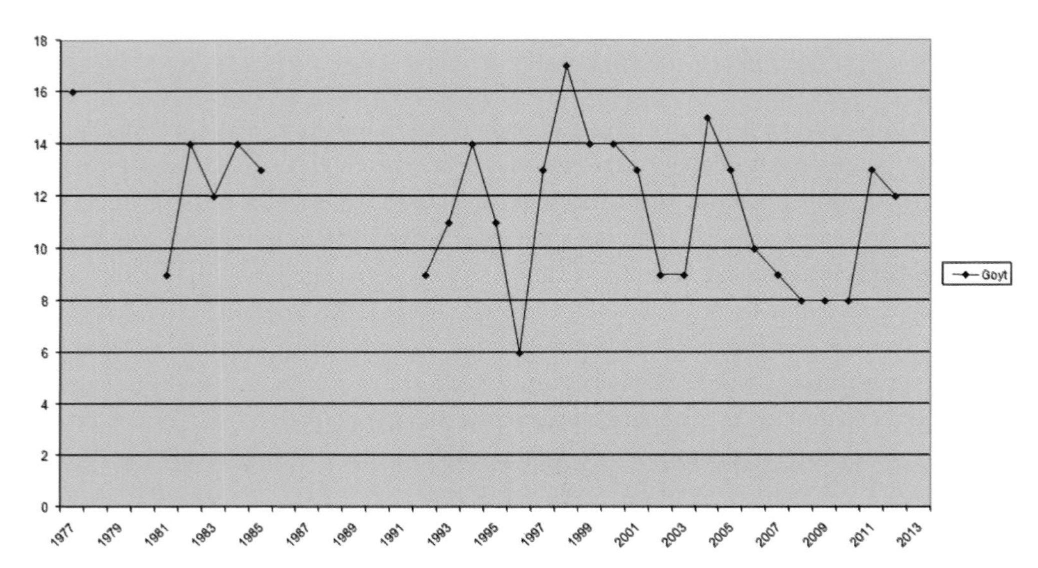

Figure 45: Goyt population.

It is thus reasonable to deduce that the reservoirs are more attractive than the streams. The probable general advantages of a lake/loch/reservoir are the sandpiper's ability to dive into reasonably deep water thus avoiding hawk and falcon predation, and also less severity in water level fluctuations during heavy rain, which reduces the risk of chicks drowning.

Another significant reservoir system in the Goyt Valley was also surveyed most years, and again the population is less variable and has not declined.

When we started our study, the feeling was that the 'natural' streams were more attractive than the 'artificial' reservoirs: another hypothesis that bit the dust.

EXPERIENCED BREEDERS AS THE NUCLEUS AROUND WHICH SUCCESS GATHERS

The 'experience' of each breeder is set at 1 when it is first colour-ringed in the Ashop (unless it has already been ringed as a chick or juvenile, in which case its age is used if different). All the colour-ringed birds thus have an experience (going up potentially to 15 in the oldest bird) and in any year the average experience is normally about 3. But if we compare 1981 and 2005, which both had similar populations, there is a massive difference in experience. In 1981 a few birds had died/deserted as a result of extreme weather, but the average experience of those remaining was 3.11 and their success was high with 17 fledged. This experienced group remained the nucleus for a steady increase in the population over the following years, and the average experience increased to reach 3.74 in 1987. But in 2005 the average experience was only 2.24, and their success was poor (only six fledged) and many did not come back again so the jump in population was temporary.

During the period of sustained growth (1981–88) there was a nucleus around which to build, and the average experience through those years was 3.56. The drop in 1989 led to a big loss of experience (47 per cent of bird-years of breeding experience disappeared). Whether this was a random occurrence in that a lot of old birds naturally reached death through old age, or whether just a few did and the others deserted, is not known but anyway the population struggled for a while and over the next seven years (1990–96) the average experience was only 2.55 and the return rate was low.

Using the same convention, the thriving LPI population had experience of 3.5 to 3.7 years (OLM). If an experienced male succeeded he had a 92 per cent return rate; if he failed, a 48 per cent return rate; and if he failed as a first attempt he only had a 20 per cent return rate (Oring & Lank 1982).

Thus returning to the question posed in this section, which was 'Can we explain the variations in the Ashop population?' the answer is partially yes. Small streams are inherently less attractive, but if they have a successful gang of experienced breeders these will produce young and they will attract recruits and they will return again. But there is slightly more attraction to the reservoirs, and if there are vacancies there and its gang fails to return to the stream, an individual will be tempted to desert. However by 2010 the reservoirs were overcrowded and it became sensible to return to the streams. The initial very high population in the 1970s of over 20 pairs was probably a group of successful birds aided by the low predation due to absence of falcons and hawks due to pesticides, and diligent control of mammals by gamekeepers. A population of 10–15 in the valley is sensibly consistent with the habitat available being used at densities similar to those of other streams in England.

However, there is no way of predicting in detail what will happen next year even if we had perfect weather forecasts and knew how many fledglings had been raised throughout the Peak District. Sandpipers are individuals not automatons, and one experienced bird failing to return can start an unpredictable chain of consequences.

LARGER RIVERS

A poster at the 2016 IWSG conference from Marek Elas on his PhD project on the River Vistula near Warsaw, where the river ranged from 400 to 900 metres in width with islands and braided channels, shows the challenges that a large river poses. In 2014 a flood destroyed every nest around 20 May, and the next year a flood destroyed 41 per cent of the nests around 1 June, so re-nesting was a big necessity. However of the nests that were not flooded 63 per cent hatched (70 per cent on islands and 51 per cent on shore) suggesting the value of islands (and floods) at reducing mammalian predators and producing attractive habitat.

In world terms Britain really does not have large rivers, but the Lune (about 30 to 80 metres wide) has been surveyed from Lancaster up to Kirby Lonsdale (32.8 km) since the start of the Waterways Bird Survey organised by the BTO, and many of the same observers and techniques are still used today. Floods are again an important variable in breeding success from year to year; winter 2015/16 had extreme floods, so it will be interesting to see the effect over the next few years.

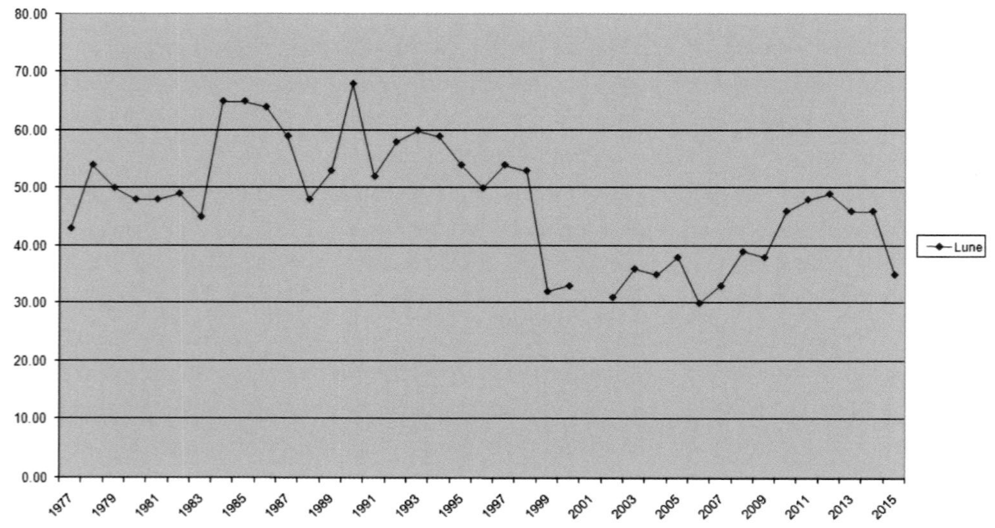

Figure 46: The variation in numbers along the River Lune from Lancaster to Kirkby Lonsdale surveyed by Lancaster and District Bird Watching Society members shows a range of variation that is similar to the LDH reservoirs, and the total number is similar for a similar length (from Annual Reports of the Birds of Lancaster & District, LDBWS).

BRITISH ISLES TRENDS FROM ATLASES AND ANNUAL BREEDING BIRD SURVEYS

Atlases at 20-year intervals monitor changes in distribution. When the first atlas was produced (Sharrock 1976) there was confirmed breeding in 1,262 of the 10 km squares, 0 in Devon at the SW extremity, and 6 on Shetland at the NE extremity. That atlas pointed out that in Ireland much of the inland habitat used was peculiar within the British Isles in being *lowland* rivers and lakes. By the third atlas (Balmer et al 2013) breeding was confirmed in only 734 squares. However the breeders were back in Devon, and still in 5 squares on Shetland. The island of Ireland's sandpiper population had declined drastically, from around 234 squares with confirmed breeding down to around 53. Most of the 'peculiar' Irish lowland squares had lost their sandpipers, but the same sort of habitat which they like in north-west Scotland, including where mountain streams feed into the indented seashores of the west, were still occupied. Indeed, in some such Irish squares the relative abundance change map showed them increasing. Of the total of Irish squares deserted, roughly 40 per cent were in lowlands. The next big loser was Wales, where the confirmed squares dropped to half. In the main part of their continuous range north from the Peak District the attrition is around the edges, and in the squares that I am reasonably familiar with there were very few birds to lose. The squares that have good habitat were all still occupied. Thus although the loss was from 42 per cent of squares, these were all peripheral and in numerical terms probably held less than 10 per cent of the population.

Reduction of range can vary from one extreme where the habitat is being destroyed (e.g. being drained and built upon) to the other where the species is no longer breeding well enough and is in general retreat (e.g. the loss of the red-backed shrike). Neither of those extremes applies here; the variability apparent in Figures 44, 45 and 46 is around a fairly steady population, though perhaps one that is only just managing and so is unable to provide a surplus for marginal areas.

The official population trend comes from annual numerical censuses which were originally the Waterways Bird Survey (WBS) that involved around nine visits to self-selected places plotting territories, but was replaced from the late 1990s by the Waterways Breeding Bird Survey (WBBS) which involves two visits to places that had been selected randomly. If people were unavailable to do the visits, the surveys did not get done. In both methodologies rivers and canals are surveyed, but lochs and reservoirs are not. From 1974 to 1988 the population reported in Marchant et al 1990 was fairly steady (±10 per cent). Later data show a steady reduction from 1985 to 2015 by 50 per cent (Robinson et al 2015), which is the primary source in bird trends (WBS+WBBS).

Many farmland and woodland species are monitored by BBS, but because all species are recorded by the observers there are enough sightings of common sandpiper to give a trend. This method started in 1995, and over 18 years showed a 36 per cent reduction in England and a 14 per cent reduction in Scotland. But most of that decrease occurred in the first half of the period.

I have some difficulty in resolving the different sources of data into a coherent picture, and especially the balance between range changes and population changes. In the core populations which we have graphed on the Lune, Goyt and LDH, it has been seen that the population decline has been negligible. The third atlas also shows many good squares in north-west Scotland as well as Ireland have actually increased their abundance; LDH increased from 2001 to 2010 by around 80 per cent. LDH has a great advantage of being a large target area on the route of most other birds going north. But the other sign that the population is reasonably healthy is the number that colonise good suitable new habitat when it is created (see Chapter 9 on habitat). The WBBS probably inadequately represents what is happening on Scottish areas (particularly lochs) and on reservoirs countrywide. Thus my interim deduction is that WBBS has a bias towards easily accessible and thus more disturbed linear watercourses, and so gives an over-pessimistic picture of the decline across the whole British Isles, though possibly right for England. But even if it was 25 per cent over the last 30 years rather than 50 per cent, it is still worrying: the retreat from marginal areas is unequivocal.

IS POPULATION DRIVEN BY EVENTS AT OTHER SEASONS?

WHEN DO THEY DIE IN THE YEAR?

If we were considering a large enough local population of birds to give an expectation of 120 of them dying per year, would we expect them to die at the rate of 10 every month? Unfortunately, for two-thirds of the year (Chapter 3) they are in areas that are very difficult to watch. Some birds must die of starvation, illness, old age and predation throughout the year, but I have found no information on anything extraordinary. The northwards migration is an attempt at optimising arrival at places as the weather improves, and is probably more stressful than just surviving the winter season. On the breeding ground there are records of spotted sandpipers killed by mink and harrier, and dying from illness and some injuries fighting for territories, though nothing extraordinary. Their rapid departure back south may imply that they believe that they are more at risk on the breeding grounds than they will be where they are returning to. There have been ringing recoveries on the southwards migration, but again nothing extraordinary, and mostly shot. My guess would be that from July to March the rate at which 120 birds would die is less than 10 per month – at say 8 per month – and from April to June the race north to breed doubles the rate to 16 per month.

WHAT ELSE COULD BE CRITICAL?

The fact that they are spread over such a large area during winter may be symptomatic that they are searching for suitable places that are somewhat limited. We appear to have some populations like the one in Iberia that are almost sedentary and a continuous spectrum up to those from Russia going 10,000 km to places where the juveniles struggle to even do a complete moult. All the winter habitats are under increasing human pressure. Thus it cannot be ruled out that winter conditions limit population. A key wintering habitat for European birds, the mangroves of West and central Africa, declined by 16 per cent from 1980 to 2005 (Spalding et al 2010), and pressures in the Sahel are high (Zwarts et al 2009).

The northwards migration is dependent (in the case of Europe) on conditions between the Sahel and the south coast of Europe. The concentration of migrating birds going across the shortest route from the Senegal and Niger rivers to the Atlas mountains and across to Spain looks critical, and bad conditions there could have annual effects and the long term climatic changes plus desertification appear to be detrimental.

SPOTTED SANDPIPER POPULATION VARIATION LOCALLY

As we saw in Chapter 1, some superb studies have been done of breeding spotted sandpipers. The most detailed study was on Little Pelican Island (LPI) on Leech Lake in Minnesota: the number of birds there can be found in Oring & Knudsen 1972 for 1972, OLM for 1973 to 1981, Colwell & Oring 1989 for 1982 to 1985 and Reed & Oring 1993 for 1986 to 1991 (though this covered a wider area, so in 1990 and 1991 I have assumed a third of the birds are not on LPI). Thereafter detailed reports ceased and the island got colonised by larger birds (Mortensen & Ringle 2007); common terns started nesting in 1989, ring-billed gulls started nesting in 1994 and reached 4576 nests by 2006, herring gulls from 1993, a pelican in 1993 and double-crested cormorants from 1998, so I have simply guessed a linear collapse from 1991 to 1994. The island is now bare, and unsuitable for spotted sandpipers. The trend is plotted in Figure 47.

Although we focus a great deal on polyandry by spotted sandpipers, many of the clutches the female lays are replacements and quite a few females only get one male, so the ratio of males to females may be less than you might expect. The variability of population is the key issue for this chapter. After a bad year causing egg loss (Maxson & Oring 1978), mice were removed from the island. The sandpiper population reached a peak in 1981

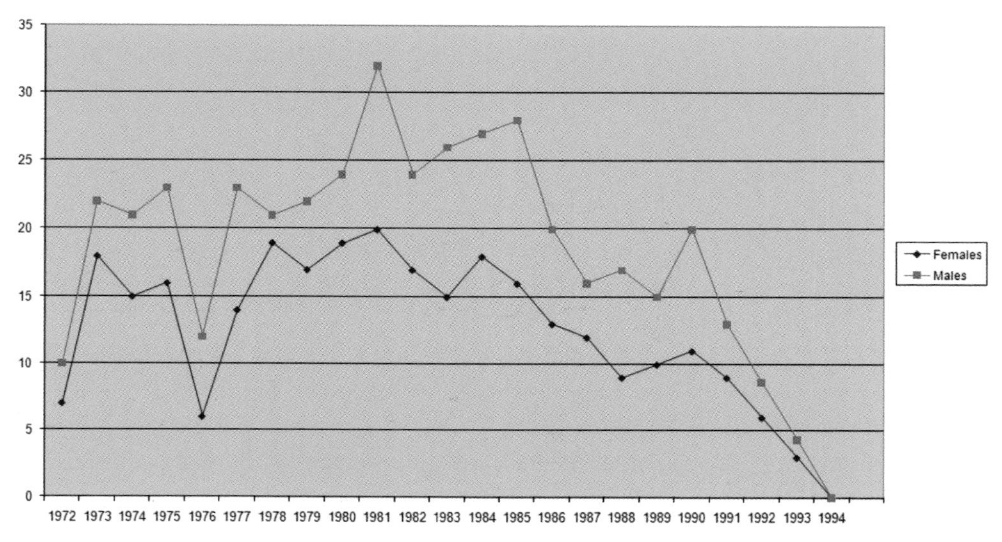

Figure 47: Number of males and females on LPI deduced from various references listed in the text, and guessing the final demise of the population after 1991.

with high numbers maintained to 1984, but then a reduction in females started. All the big birds that eventually took over LPI did not, presumably, suddenly arrive in the 1990s, but would have been building up on neighbouring areas, thus acting as a deterrent to sandpiper breeding. Leech Lake is primarily a fishing and recreational area for people in May and June. The ornithologists in towers on the island were not seen as a threat by the sandpipers but may have been a slight deterrent to human and other visitors. Removing mice and mink was a good thing for sandpipers. Thus the reduction of threats through the 1970s and early 80s helped – but then the invasion of big birds in the 1990s destroyed this heaven for spotted sandpipers. Incidentally one of the males which bred in 1986 had been reared in captivity (Pickett et al 1989); he deserted both of the clutches that were laid for him (this was quite abnormal as only 7.6 per cent of clutches were deserted in the LPI study in spite of the degree of aggressive behaviour going on) and he did not return in 1987.

The situation on Gull Island off New York that Helen Hays had reported on in 1972 (Hays 1972) has remained approximately the same (Hays priv. comm.). Terns are still conserved, and whatever is done to support them also keeps spotted sandpipers happy.

IN CONCLUSION

There is good reason to believe the numbers of sandpipers have been decreasing over the last 50 years. Their flexibility makes accurate censuses over large areas quite difficult, and detailed studies have been done in areas more strongly influenced by people than the sandpipers' core breeding areas. In the last 50 years the number of people has doubled, and in addition their ability (per person) to disturb or destroy waterside habitat has increased markedly, so decline is probably inevitable given present human behaviours. The next chapter tries to put this in context.

8 LONGER-TERM HISTORY OF COMMON SANDPIPERS

Much of the work informing this book has been done in the north-west of England, and here I look at the possible changes in the distribution of common sandpipers that have occurred there over a long period of time. This area has seen much change in its social geography in recent centuries, but this may also be representative of other places. Well before that, it was recovering from an ice age, and this naturally leads back to where the birds came from right the way back to their origin as a species. The discussion obviously becomes ever more hypothetical going further back into history.

The focal area used in this exercise is shown in Figure 48. Each of the main rivers that are considered – Derwent, Mersey, Ribble and Lune start at around 600 metres above sea level in the Pennine Hills and have broadly similar catchments in area and geomorphology

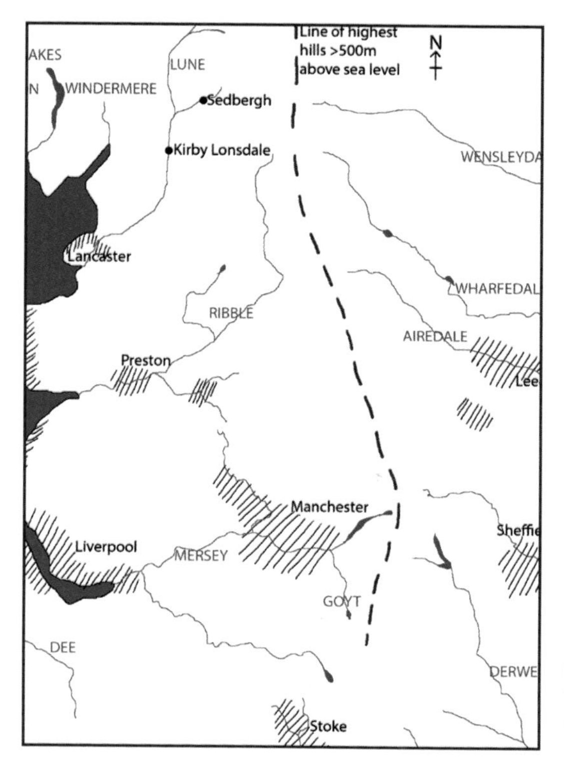

Figure 48: This sketched map shows the places mentioned in this chapter, which considers the many changes which affect common sandpipers. The focal area includes the counties of Cheshire and Lancashire and their associated cities of Liverpool, Manchester, Lancaster and Preston, the part of Derbyshire that makes up the Peak District, and parts of Yorkshire and Cumbria. No sandpipers regularly breed further south in England and probably never have, as the lowland rivers are essentially large drainage ditches with densely vegetated banks that in times of high rain respond by overflowing into meadows rather than flowing extremely fast and eroding the banks. All the streams and rivers with sandpipers start in the Pennine Hills that form the 'backbone' of northern England at over 500 metres above sea level. This rectangle has about half the English population of common sandpipers (though only a small part of the population of the British Isles, because most are in Scotland).

1950–PRESENT

A general decline during this period has been covered in the previous chapter.

1850–1950

In the latter years of 19th-century Britain there was a surge of books about the birds of individual counties. Victorians were also great collectors of eggs, stuffed birds and other creatures. Thus there is a range of sources to produce a picture of where sandpipers were found. Many natural history societies started during this period and indeed the Royal Society for the Protection of Birds had roots in the Manchester area in the 1890s as Great Crested Grebes (*Podiceps cristatus*) were plunging towards local extinction due to the desirability of their feathers to decorate hats for fashionable ladies.

The cities that had been very densely populated slums steadily expanded into big suburban areas, reservoirs were built, river courses were altered for industry and housing (mainly the Mersey and the construction of the Manchester Ship Canal by 1900), and many pollutants were increasing. Two major wars and the great depression affected the area and agriculture changed from highly labour intensive in 1850 to highly mechanised by 1950.

Such as 'climate change' was an issue at all, there was a climatic warming starting from around 1850 and going through towards 1950 though by the middle of the 20th century there was some concern that we may be drifting into a new ice age but soon the reality of man-made acceleration of warming dawned. In Burton's review of Climate Change and Birds (Burton 1995) our sandpiper is judged to have been unaffected.

So overall what was happening to sandpipers? They were often known as 'summer snipe' at the start of this period but this term soon completely disappeared. The county books – Lancashire (Mitchell 1885), Cheshire (Coward 1910), Derbyshire (Whitlock 1893) – mention their presence along the main rivers and up the streams. There were few reservoirs so they do not feature as habitat. Words like abundant and widespread are used and the reasonable interpretation of people looking at longer term trends (eg Holloway 1996) is that they were less abundant in the 1970s than in the late 19th century. Johns 1910 wrote '…one need only repair to a retired district abounding in lakes and streams and in all probability this lively bird will be found….'. In modern speech the adjective 'retired' would mean away from the towns and habitations.

The general picture from 1850 to 1950 in northern England is of a commoner bird than today.

1750–1850

This interval starts with our countryside having the first stirrings of the Industrial Revolution with massive factory innovation on the lower parts of the Derwent by Richard Arkwright at his mill at Cromford, canals by the Duke of Bridgwater around Manchester to transport coal and Samuel Crompton inventing his spinning mule in Lancashire. The revolution progresses with the proliferation of railways following on the success of the Liverpool and Manchester Railway in 1830. The movement of people into overcrowded cities led to Frederich Engels writing about the condition of the working classes of Manchester in 1844 and helped to launch Communism. Many people had left the countryside and I have thought, when sat on a pile of stones that was once a building watching sandpipers, that 200 years ago there would have been busy people, dogs, cats, chickens and probably too much disturbance for sandpipers to breed successfully there.

Near the beginning of the period Gilbert White writing from his home at Selborne in the south of England in his 'letter' of 8 October 1768 about an unexpected local breeding record of common sandpiper commented that sandpipers usually only bred 'up north' which is at its simplest level how its distribution would be described by southern birdwatchers today. Interestingly he did not mention Wales or the West Country although many of his correspondents were from those areas. The stained glass window celebrating Gilbert White in Selborne church includes a common sandpiper feeding at the feet of St Francis of Assisi.

William Wordsworth who lived in the Lake District in the northwest corner of Figure 48 recorded one in his 1789 poem which has a habitat description in its lengthy title – 'Lines left upon a Seat in a Yew-Tree which stands near the Lake of Esthwaite, on a desolate part of the Shore, commanding a beautiful Prospect'

> … here he loved to sit
> His only visitants, a straggling sheep,
> The stone-chat, or the glancing sand-piper;

Thomas Bewick produced his *History of British Birds (Part 2)* in 1804. He was born in 1763 and spent his childhood on the banks of the Tyne (just north of the focal area, but a stronghold for sandpipers in England). He must have been familiar with all the river's

Figure 49: In Saint Mary's church at Selborne, the Rev. Gilbert White memorial window (by Alexander Gascoyne 1920) is probably the only common sandpiper celebration in stained glass anywhere.

birds, as his biographer (Uglow 2006) says 'we can say for certain as a boy he was down by the river from spring until autumn'. Yet in his book he describes the sandpiper as 'not numerous ... pebbly brooks and rivers ... lays 5 eggs ... leave in Autumn'. For a boy who spent all his boyhood on the banks of the Tyne near where sandpipers breed today and did not remember nests with four eggs is surprising. His book is illustrated with his famous 'woodcut' engravings and he needed to get his specimen bird from a landowner 15 miles away on the River Wansbeck, this hardly gives evidence that the species was very common; probably it was less common than it is now as many people lived and worked along the rivers. The landowner himself was an interesting character at Little Harle Tower. When he married at 21 he changed his name from the Right Honourable and Reverend Lord Charles Murray (son of a duke) taking his wife's name of Aynesly, but by the time that he died 16 years later he was the Dean of Bocking in Essex (500 km away) while his wife was spending money extending the house at Little Harle (now a listed historical site). There must be a story there, but not for a book about sandpipers! The post of Dean of Bocking is itself suitable material for Bill Bryson; the parish was given in 995 to the Archbishop of Canterbury who still appoints the dean 1,000 years later, though the Bocking church does not really serve much purpose so the dean actually serves at a church in another town.

The end of what was known as the Little Ice Age happened at the end of this period as weather became less severe (though this was mainly winter weather anyway so its impact on birds migrating in to breed in May would have been small). In 1822 a list of first dates for migrants in Manchester gives April 29 for sandpiper much as it might be there today. But it is a later date than an attractive area for breeding where the birds arrive earlier.

My general impression of 1750 to 1850 is of a bird no more common than today, and possibly even less common.

TWO AND A HALF CENTURIES OF CHANGE

In the above brief history we have mentioned several factors that have changed. Taking each in turn and focussing mostly on the catchment of the River Mersey:

URBANISATION AND SUBURBANISATION

Undoubtedly some good habitat had already been lost by pre-1850 industrial change. However the mass movement of people from the countryside to the cities during the 19th century will have reduced the numbers of people and their animals around sandpiper habitat. Then in the 20th century lower density housing was built outside of cities for ever more people, so the homes for sandpipers were taken away. All the rivers that are tributaries of the Mersey (Thame, Goyt, Irwell) became quite full of businesses and houses during the 20th century. Nevertheless in 1930 in a survey of the birds of Heywood (Whitaker 1932), the author can still report six territories of sandpipers along a stream that by 1980 had none, and report 20 pairs in an area where only the occasional pair breed now as housing and recreational areas have continued its suburbanisation. Nationwide from 1750 to now the human population grew roughly ten times, though much has been in cities where sandpipers did not breed.

RESERVOIRS

These created new potential habitat along their banks. The streams and rivers that had flowed there before the dams were built were largely steep-sided and more suitable for dippers than common sandpipers. In the Mersey catchment by 1970 every upland tributary had a reservoir. The growth in reservoir edge just in the Mersey catchment shows at least 56 km being provided mainly in two periods. Firstly from 1850 to 1880 when the City of Manchester developed the Longdendale valley (River Etherow) as a string of five reservoirs to provide controlled flow to factories as well as drinking water. Latterly from 1960 to 1970 there was another surge of building reservoirs. Mersey tributary reservoirs provided 40 of the territories in the review of the total population of the Peak District (Holland et al 1982b). Together with the 60 around the main Derwent Valley reservoirs and another 20 or so on small reservoirs, most of the Peak District birds nowadays (and increasingly so ever since 1850) are using a man-made habitat that did not exist 200 years ago.

In the particular environment of the Derwent reservoirs where 33 km of reservoir bank now exists which has around 60 pairs (Figure 44), a look at the pre-reservoir maps suggests there may well have been only 20 pairs along the natural streams. It is fair to describe the boom of reservoir construction in this part of England as a large habitat creation exercise for common sandpipers (but bad news for dippers).

Figure 50: Upstream of Yorkshire Bridge, the River Derwent splits into two. Before reservoirs the two catchments were similar upland farming with similar size (small) rivers, and so should reasonably be expected to have similar numbers of sandpipers. Nowadays the catchment with reservoirs has considerably more (Figure 44); the reservoir banks are unstable and the undercut bank provides good hiding and feeding places for chicks as well as leaving a pebbly shore.

CHANGES DOWNSTREAM

Creation of the reservoirs in the headwaters and other engineering in the lowlands of the River Mersey reduced the power of the flow to produce the eroding bends and shingle banks lower down the river that would attract breeding sandpipers. By contrast the River Lune is still fairly natural and has around 40 territories in the 32.8 km stretch from Kirby Lonsdale to Lancaster. The Mersey now has none in an equivalent length where Coward in 1910 described them as widespread breeders. The mill that Coward's father owned by the Mersey just downstream from Stockport was in an area that I did a Common Bird Census in the 1970s. Sandpipers still visited on spring passage but the river there now has a flood protection bank that prevented creation of shingle edge and a footpath along the bank with a lot of dogs and their walkers. From 1880 to 1980 the locality changed from a farming village with a mill and a new railway station to total suburbia.

The situation lower down the Mersey is further complicated by the development of 'flashes'. These are areas where the ground has subsided due to underground coal or salt extraction. When new, these flashes can be good habitat especially if the banks are industrial spoil that is essentially the same as a shingle bank. There was thus a period when the degree of reservoir development had not totally reduced flows (so there was still some naturalness in the main river) and new flashes were developing. Boyd (1951) in his book 'A Country Parish' describing the wildlife around Great Budworth still reported a few sandpipers, but by the time of the first Cheshire Bird Atlas (Guest et al 1992) they were gone. Pennington Flash appeared due to subsidence around 1900 and had several breeding sandpipers in the 1940s but they ceased to breed there by 1968 (Wilson 1985). Increasing vegetation plus recreation pressures are the attributed cause.

The Ribble and Lune flows have not been so affected by reservoirs and also they are less affected by urbanisation. Blackburn, Burnley and other towns took up valleys for human activity up some tributaries of the Ribble and a large reservoir up one of the major tributaries (Stocks reservoir up the Hodder) has around 15 sandpiper pairs. The Lune remains rural and far enough from large cities and has no reservoirs. The lower Lune population was shown in Figure 46.

POLLUTION

The rivers of the area have varied histories. The Lune had a few factories in the tidal region but very few upstream or on any tributary and it remained a reasonable salmon and trout river throughout. The Ribble and its tributaries were more industrial. Parts of the Mersey became a case of near-total death by industrial and sewage effluent but the effect on sandpipers is not easily separable from the other human impacts just discussed. However there must have been sites close to effluent discharges that were completely useless for sandpipers.

When we were starting our studies of sandpipers, the Peak District was used by the Central Electricity Generating Board as a study area for the effect of 'acid rain' due to being surrounded by coal fired power stations. Their studies found good numbers of stream invertebrates in the Ashop tributaries where sandpipers occurred. Contrarily a decline in common sandpipers on streams in the Peak District was roughly while flue gas desulphurisation was fitted to the surrounding power stations but this is unlikely to be a reason for their decline. Studies elsewhere (Vickery 1991) suggested that stream acidification was not a factor in sandpiper ecology (though it was for dippers).

During the 1990s the chemical used to protect sheep (sheep-dip) was changed from one hazardous for people to synthetic pyrethrins but these were soon (by 2006) banned due to the effect on freshwater invertebrates. This could have had a localised effect on chick survival for a while in sheep districts.

AGRICULTURE AND FORESTRY

With mechanised farming the arable landscape has changed enormously but the sheep and cattle and associated hay of the upland riverside fields has been less affected. Less work-horses allows more sheep or cattle but evidence that one creature has more impact than another has not been put forward. Tom Dougall (priv.comm.) noted that when sheep grazing was terminated the bank vegetation got denser and the sandpipers got scarcer. The introduction of chemicals as sheep dips, fertilisers or herbicides was low impact apparently. The change to synthetic pyrethrins for sheep dipping has been mentioned under pollution.

Following the 1914–18 war there was a push to increase acreage of forests above the very low level that had been reached and this push continued. This led to a large increase in forests of conifers in the uplands. When done right to the stream edge this may have an effect and this may change with the age of the planting. Thus overall this is probably just another temporary potential detriment comparable with the other impacts after 1950. In Russia and other forested countries the common sandpiper is seen as a bird of forests and its world distribution largely aligns with boreal forest. Nests are found well away from water among the trees as they also sometimes are around forested Peak District reservoirs.

CLIMATE CHANGE

Burton 1995 reckoned the effect from 1850 to 1950 to be negligible. If there is an overall detriment from recent man-made climate change it has not been detectable relative to the local changes in land use.

GAMEKEEPING

According to census returns the peak of gamekeeping occurred in 1911 having increased to 23,056 people so employed (Tapper 1992). Grouse shooting in particular was a new social imperative for the upper and upwardly mobile classes, and to ensure plenty of grouse all animals seen as vermin were heavily controlled. This reduction in predators will have helped the sandpiper on many of its breeding areas and if this led to breeding success going up from replacement rate to a little above replacement rate this would have helped the population expand. World wars sent many of the men that had been employed in killing vermin to kill people instead. As well as the number of keepers dropping to

4,391 by 1951 some of the gamekeepers' practices became illegal during the 20th century and also new vermin, particularly American Mink released from fur farms after 1970, started to spread too. We have seen in Chapter 6 that lowered predator numbers would have been helpful.

RECREATIONAL DISTURBANCE

We live in an age of leisure and tourism. Studies which showed how birds interacted with anglers and other recreational disturbance have also been described in Chapter 6. Water companies used to be keen to exclude people from their catchments but with improved water treatment and increasing demand from the public for access more water-side habitats are now available for people and less for sandpipers.

Up till the 1970s in England there were two 'bank holidays' in the sandpiper breeding season Easter and Whitsun. These were both moveable feasts according to the Christian calendar and in most years Easter is before the birds arrive and Whitsun could be any time between mid-May and mid-June. Now we have an extra holiday in first weekend of May (May Day) and Whitsun has been replaced by a 'Spring' holiday fixed on the last weekend of May; both crucial times for sandpipers and now occurring at the same time every year.

ALL EFFECTS TOGETHER

Overall the picture appears to be that local causes had local effects but the sandpiper was pretty widespread and if it disappeared from some human dominated areas it had new habitats like reservoir edges and upland streams with predators controlled by gamekeepers to move to. In Table 8.1 below is an attempt to estimate for the Mersey catchment (as the most changing catchment) where and how many common sandpiper pairs there might have been at 50 year intervals.

Table 8.1 A rough guess at how the changing Mersey catchment affected common sandpipers

	2000	1950	1900	1850	1800
Small Streams	5	10	35	30	25
Tributary rivers	0	5	10	15	20
Main river	0	5	15	20	30
Reservoirs	30	35	20	5	0
Meres and flashes	5	15	20	10	10
Total	40	70	100	80	85

It is emphasised that these are general judgements (opinions) of mine based on the factors outlined above.

Thus for streams the numbers increase slightly as subsistence agricultural workers leave for factories but decrease as the valleys are flooded for reservoirs, taken over by housing and used for recreation. The lower river valleys get more houses and factories. The lower Mersey loses flood spates due to reservoir construction and embankment flood defences. New reservoirs create new habitat but later get recreational pressure. Meres and flashes may sometimes have good habitat but with age get vegetated edges and taken over for recreation. Overall pollution rises and then falls and so does game-keeping.

When I started researching this chapter I had thought that the sandpiper decline had been going on for ever as human population expanded. However I have come to believe that the late 19th century and early 20th century may have been a time of high sandpiper population after a period of new habitat creation, extreme predator control and people too busy working in factories and warring overseas to disturb them.

IRON AGE TO 1750

The state of sandpipers in the northwest of England can only be inferred from the very scanty clues that there may be about their presence in the British Isles as a whole.

In areas that have sandpipers today their arrival as a sign of spring is no less noticeable than the cuckoo or the swallow. Their record however appears to be totally absent from literature. Shakespeare, who is a great user of birds and a book has been written about the 50 species he invoked (Goodfellow 1983), made no mention of sandpipers although snipe in Othello and wagtail in King Lear were used. It may be that Shakespeare's sojourn in London made him sentimental about birds but the Avon at Stratford is not good sandpiper habitat so his omission of them is not serious. They were clearly present at that time as they are in the first list of birds given for the country by William Turner in *Avium Praecipuaram* in 1544 where his list was of 105 wild birds (Fisher 1966). The excellent thing about this list is that Turner was from Morpeth in Northumberland (though he later lived in Kew where he had a botanic garden and Wells where he was Dean until Queen Mary threw him out for protestant activities). His book is largely trying to see which birds described by Aristotle and Pliny he has seen himself in Britain so he adds his own observations – on dippers ('black and white, short tail, dips repeatedly, I have observed it on the banks of streams but nowhere else') and common sandpipers ('black on back, white on belly, long shanks and a bill by no means short; in spring it is exceeding clamorous and querulous about the banks of rivers'). The quotations are from the translation by A.H. Evans, 1903, CUP.

So it would be reasonable to say it was in the top 100 most widespread birds. They were clearly about in John Ray and Francis Willoughby's time around 1650 as he describes – 'They frequent rivers and pools of water. I have seen them on the R Tame in Warwickshire and Lake Geneva'. That Ray invoked Lake Geneva rather than somewhere in northern Britain seems odd when for grey wagtail he says that 'they frequent stony rivers and feed upon water insects' and for dippers he says 'frequents stony rivers in the mountainous parts of Wales, Northumberland, Yorkshire'. So this is hardly implies a very familiar breeding bird.

Artists generally included more exotic birds like peacocks and parrots to adorn their paintings of the aristocracy, or game-birds in the shooting genre of art. So ducks are arty but small shorebirds are not. In a classic painting shown in The Bird in Art (Bugler 2012) of 'The Creation of the Birds and Fishes' by van Oosten, God is standing on the shore but he did not create any shorebirds! But looking through bird artists' work for sale does not lead to a vast choice of sandpiper pictures even today. The one that I have I commissioned from Frederick J Watson so it includes a colour ring (Figure 51).

Figure 51: This painting is by Frederick J. Watson, a fine artist of seabirds, from his gallery at St Abbs in Scotland.

In the list of bones found at archaeological sites in *The History of British Birds* (Yalden & Albarella 2009), the common sandpiper is a species with no record at all. The sources of old records were the skeletal remains at various predation sites (predation by birds and mammals including Homo sapiens). The possible reasons may be

- it is only here in summer – but barn swallow is recorded 22 times along with 20 other summer visitors
- upland streams are not well covered – but there are grey wagtails, pied wagtails and dippers recorded
- small shorebirds are not well covered – but there are 10 green sandpipers, 8 common ringed plovers and 18 dunlin
- Scottish lochs are not sampled – but there are 36 divers (loons) and 11 greenshanks'

One possible reason for this absence of remains is that they are present for such a short time when other prey is abundant and more easily caught. Clearly another possibility is that they were relatively scarcer in the past than now.

From Old English (OE) place names, Yalden and Albarella 2009 give a list of bird-based names. The popular ones are derived from eagles, hawks, cranes and crows, but two are reckoned to be from 'stint' which was the OE word for 'sandpipers'. One is Stinchcombe in Gloucestershire, which is by a stream draining off the Cotswold Hills (combe being OE for valley); the other is Stinsford in Dorset, where a ford crosses a stream. My judgement would be (assuming that the 'stin' does refer to a bird rather than something else) that both these places were regular haunts for green sandpiper from August to April and possibly passage sites for commons in May and July. A spot check of a recent Gloucestershire bird report (2006) had more records of green than common.

My guess is that in olden times the common sandpiper was not exceedingly more common in northern England than today.

ICE AGE TO IRON AGE

The land was covered with ice about 12,000 years ago. The sea level was lower and the waves met a barren land. The gradual increase of people into the area as the climate improved occurred along with their developing social organisation and technological advances. For this book the point of interest is did the sandpiper arrive well before us, with us or slowly after us?

In their current distribution they do not push hard against the tundra while humans do. Sandpipers reach just about a mid-May 10°C isotherm. As a species that has low breeding productivity I would guess that as the ice retreated they spread less quickly than we did.

Through the previous hundreds of thousands of years of ice sheets growing and contracting the range of the common sandpiper will have followed the available habitat. Ideas about the refuges around the Adriatic and Black Seas that redshank used during the periods of maximum ice (Hale 1980) may reasonably apply to our sandpipers basically retreating as breeders to available habitat further south. Along the stretch of the River Lune where I did a Waterways Bird Survey, redshank and common sandpiper lived side by side in similar numbers. The fact that redshanks have reached Iceland but common sandpipers have not may be suggestive that the common sandpiper arrival in the west of the British Isles is more recent than the arrival of redshanks.

SPOTTED SANDPIPER {*ACTITIS MACULARIUS*}

To an inhabitant of Manchester (England) reading of spotted sandpipers on an island in a large lake in rural Minnesota is a dream of unspoilt wilderness. However, in reality there are some similarities to the Peak District. Leech Lake's level was increased by around 2 metres by a dam in 1882 which caused several lakes to combine and bits of land to be divided into new islands. The railway came in 1896 leading to development of tourism and all the things which that implies on a lake; boating, fishing and beach-side properties. The lake is stocked with fish particularly Walleye. Cormorants are controlled and people are warned not to eat too many fish because of Mercury contamination. As we saw in the previous chapter, by the 1990s the island had ceased being habitable for sandpipers. The name Pelican Island presumably means that before they were all shot the place was more suitable for them than sandpipers. Overall, man's impact is clear.

The highest point in all Minnesota is 742 metres and those around Leech Lake are much the same height as the Pennines. The outlet feeds a tributary in the headwaters of the Mississippi. In one paper (Reed et al 1996) the next place with a decent population of spotted sandpipers is said to be 225 km away to the north in the Lake of the Woods and the recent map of the distribution density across the country in BNA shows a low density in a large area to the west of Lake Superior. It may thus be that LPI was quite an exceptional place. Making sense of the general changing distribution in America is beyond me but many of the factors affecting common sandpipers in the north of England will presumably be in action.

BACK IN TIME TO ORIGINS OF THE *ACTITIS GENUS*

The estimate of when *Actitis* became distinct from their closest relatives like redshanks is about 35 million years ago (Mya). This estimate is taken from Baker et al 2007 who did a detailed study of 90 shorebird genera looking at their DNA. The gradual divergence of a family tree from a common ancestor of shorebirds (that is quoted as having arrived around 93 million years ago) is shown in simplified form in Figure 52. This family tree from DNA is not widely different from that which had been used for decades in bird lists based on anatomy and behaviours. The uniformity of common sandpipers through their range as observed by birders and in museum specimens is confirmed by DNA sampling covering the whole of Russia from 34deg E to 152degE which indicated a single lineage. The divergence between spotted and common is only 12.6 per cent in the same study (Zink et al 2008). Within America, DNA comparison also showed that spotted sandpiper has no variation from one place to another (Reed et al 1996).

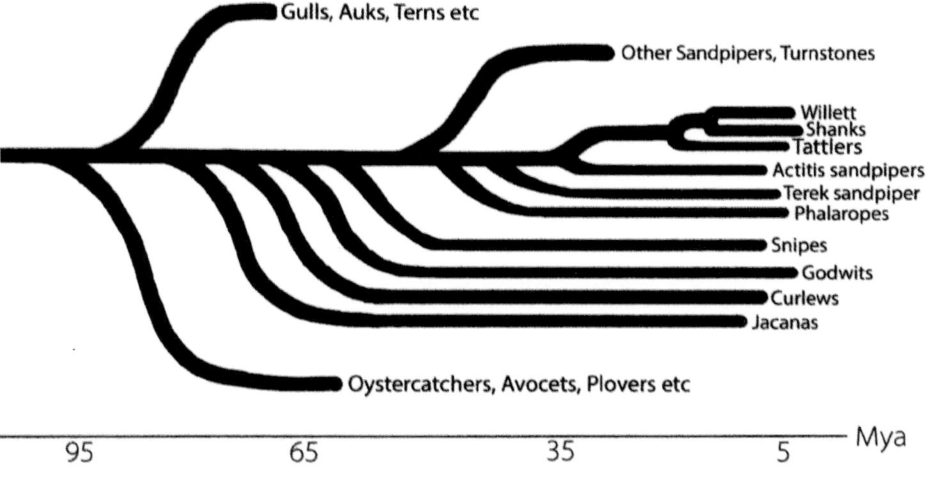

Figure 52: The key branches in evolution of various shorebirds simplified, from Baker et al 2007.

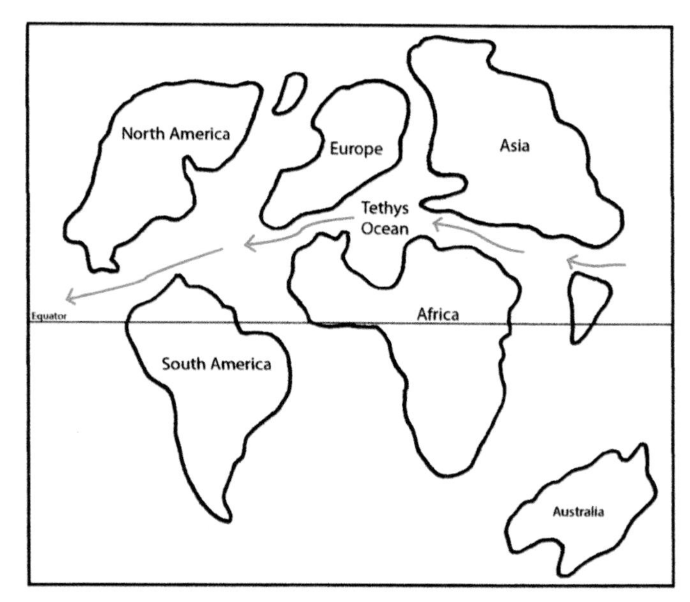

Figure 53: The continents around the period that *Actitis* was established. Because there were gaps between southern and northern continents, ocean currents were completely different. The world was warmer by 2–3°C, and the sea level much higher.

The occurrence of lakes and streams with gravelly edges were widespread as the continents drifted and rain patterns changed. The world of 35 Mya had the continents in different layout than present (see Figure 53), the sea level relative to now was much higher (but dropping) and the temperature warmer (and also dropping). Studies of today's rocks that were then deep-sea sediments show that big rivers were pushing much sediment into the sea and these will have started, as today, in many small streams from inland uplands and so the habitats and food which we associate with breeding sandpipers today were there to be exploited. In Chapter 3 we saw how in January the dense concentrations of both *Actitis* species are now in mangrove forests. Mangroves were well established by 35Mya and were prospering on the edges of the continents around the Tethys Ocean where it would be reasonable to deduce that *Actitis* evolved its distinctive life habits.

So *Actitis* could exploit the mangrove habitat, and migrate up from there to use a breeding habitat on the edges of rivers and lakes at times of year with high food abundance there. When the non-breeding and breeding places got further apart and the climate changed, its migration capability could adapt in response. In a warm and wet world it would be easy to spread from catchment to catchment, and with the distance from Europe to Greenland and then North America being only two crossings of about 500 km at that time, their range including all the northern temperate lands is entirely reasonable. As a general purpose small shorebird it could fill a fairly universal niche where seasonal high precipitation kept a waterside free of too much vegetation which then became exposed in warmer and drier season, so it was good for insects and other similar food. Their divergence of breeding tactics into polyandry on one continent and monogamy on the others is looked at in the next chapter.

OTHER COMMON SANDPIPERS

Once large icy regions appeared that were only habitable for a very short time in summer, these needed to be colonised by species better adapted for very long migrations and the use of open habitat for large flocks to gather when not breeding. The *Calidris* sandpipers took up this challenge and diversified into Baird's sandpipers, broad-billed sandpipers, curlew sandpipers, dunlin (still diverging with many subspecies), knots (with many sub-species), least sandpipers, little stints, long-toed stints, pectoral sandpipers, purple sandpipers, red-necked stints, rock sandpipers, sanderlings, sharp-tailed sandpiper, surfbirds, western sandpipers, white-rumped sandpipers … with their amazing migrations and specialised exploitations of habitats and totalling about 20 million birds. And at the most extreme the spoon-billed sandpiper at some stage in this explosion of diversity found its peculiar bill an evolutionary advantage. None of these is routinely capable of multi-clutching, though some females lay two quick clutches: one for herself and the other for her male. Wood, green, solitary sandpipers adopted more boggy areas, and some decided to use old tree nests of other species and to winter in fresher water than *Actitis*.

Though hardly meeting the criterion of 'common' the tattlers appear to be most closely overlapping in requirements with *Actitis* and are also closely related, but are restricted to quite small ranges either side of the Bering Sea. Another close relation, the Terek sandpiper, is similarly restricted in range so these species were not quite as adaptable as *Actitis*.

9 THEIR PLACE IN THE WORLD

WHY THE DIFFERENT BREEDING STRATEGIES?

Birds have two basic choices

- to lay eggs onto something that is already there OR to build a complex nest
- to keep the chicks at that place until they can fly OR have them mobile from hatching

Shorebirds fall into those who do not build a complex nest and whose chicks move off very soon after hatching. To be strong and have vision at hatching the eggs are at the upper end of egg sizes (relative to the bird) and while resting on the ground they have a wider spectrum of predators to eat this nice meal – all mammals and snakes rather than just agile tree climbers. They may also be trodden on. Thus their breeding behaviours need to respond to this fact. One strategy is to fly to a remote and harsh area where predators are scarcer and a brief abundance of excellent breeding opportunities allows shorebirds to swamp predator pressure. Another strategy is to have very well camouflaged sites. Another strategy is to be communally aggressive towards predators. Another is the ability to lay replacement clutches. These are not mutually exclusive strategies and so a wader may fly to the Arctic, have an extremely cryptic nest, respond to the sight of a predator vigorously together with all others locally so distracting the predator significantly. But if they are the unlucky pair needing to lay a replacement clutch the season may be too short for replacement to be a major component of the strategy.

Actitis have filled a widespread habitat of temperate northern regions which abound with predators. Their nests are average in difficulty of finding and the clutch is 35–50 grams of nutrition for the finder. There is some benefit in nesting close enough to another pair so that their alarms are useful to neighbours and some joint alarming behaviour is beneficial but this is a small part of their strategic armoury. The ability to lay replacement clutches is crucial to them as also is to select places with below average predation rather than above average predation using their experience and intelligence. So it is a necessity for them to observe where they as individuals have been successful and, if they have been unsuccessful, where others of their species have been successful.

With a high likelihood of needing to lay a replacement clutch it is important that the female gets into good condition on completing a first clutch and the male does a share of incubation. Since success in any year is a lottery it is also important for the female to survive to the next year. Thus in both species the male is the main incubator and the female keeps in good condition and leaves for sunnier climes when the last brood is half grown.

So in this general strategy a female has laid her first clutch and is ready for laying a replacement. But if the first clutch is still intact she is capable of laying for another male if one turns up. If males have seen in previous years that this is a low predator place they will know that they have a good chance if they join the queue (wait in line). This practice will also gradually lead to the male only needing to have quite a small territory as long as food is superabundant.

If the habitat has quite extreme variations in quality (both food and predator abundance) then it becomes very important for the female to grab her part of it otherwise she will waste her effort laying eggs for a no-hoper male in poor habitat.

As far as the study areas are concerned there is a clear difference in the extremes of choices.

In the Peak District there are no islands of extremely low predation with extremely high food habitat. Individuals may easily move from one place to another. I suggest that reservoirs are slightly better than streams and wider rivers are slightly better than narrow streams but there is not a step change to make it worthwhile either sex queuing for one rather than the other. On a couple of cases there appears to have been an unattached male and there was the possibility that a neighbouring female laid him a clutch but that is unusual. In general, birds without a mate go and look elsewhere rather than wait about. Furthermore, from the Peak District all the way up the hills of the north through to Scotland and on to the northernmost coast of Scotland is a continuous stretch of potential territories of general similarity.

On Leech Lake there appear to have been extremes of choice. The island was extremely good with food coming from all directions and people controlling predators, competitors and disturbance. This made it a step change better than surrounding areas and worth a female fighting for and males to wait in line. Along the 0.75 km shore of this tiny island there appears to have been a comparable number of birds to the 300 km of shore around the edge of the lake. The next area with a flourishing population was reported to be 225 km away in Lake of the Woods (Reed et al 1996).

The downside of polyandry for the females appears to be they may stress themselves so much their life is shortened (or possibly they moved elsewhere when they got too old for the stress of fighting for a patch on LPI). The small studies done on spotted sandpipers on an 'ordinary' site (Oring & Knudson 1972) showed poor breeding success and little or no polyandry. I questioned at the beginning of Chapter 7 the apparent low numbers of spotted sandpipers in the world (reportedly less than the number of commons in Finland alone). Perhaps the number of adequately productive sites is very limited and the population can only sustain itself by being polyandrous when a good site is found.

PREVIOUS SUGGESTIONS ON POLYANDRY ORIGINS

As the diversity of mating systems became better known in the 1970s various theories were proposed and the evolution of polyandry led to many suggestions. These were reviewed under five headings by Erckmann in the 'Evolution of Polyandry in Shorebirds' a paper in a collection called the Social Behaviour of Female Vertebrates (Wasser 1983) which ranged from Women in Taiwan to Elephants. The headings for how shorebirds (which included jacanas) may have evolved polyandry were –

replacement clutch hypothesis – which was similar to the pragmatic description given above. The female can produce more young by producing more eggs and it advantageous to lay for multiple males AS LONG AS the males are capable of raising the brood alone. He showed that in general for small shorebirds the male was better than the female as a single parent. In a population of western sandpipers (which is monogamous) he collected 12 females thus leaving the male as single parents and 12 males from other pairs thus leaving the female as single parent. The females gave up after four days while the males kept going for eight days. The harsh Arctic conditions where they bred clearly needed both of the pair to keep the eggs warm and to have enough time to feed but the male was better at it. But it does emphasise that the male needs to have the single parent incubation capability in its environment for polyandry to work and in high food, warm, continental climate the male spotted sandpiper clearly is capable.

evolution from double clutching – this proposed that the first step was to lay one clutch for the male and one for herself (like Temminck's stint, little stint and sanderling have been seen to do). This was judged unlikely as the few species that do this are not closely related to *Actitis* and that behaviour probably evolved separately rather than a step on the road to polyandry. Another observation is that although the female spotted sandpiper helps with her last brood (as of course does the common) the female in double clutching species has no involvement whatsoever with the male's brood once she lays one for herself.

stressed female – this suggested that the female was so stressed from laying four large eggs that it had no energy left to incubate and so the male had to do it. When she recovered she started with a new male. This was tested in various ways for the various species showing polyandry and was dismissed. The thought that one day the female is too stressed to incubate and yet a day later is fighting off rivals and soon laying another clutch is intrinsically bizarre for spotted sandpipers.

fluctuating food – this idea was that when food was abundant the female should grab the opportunity to lay multiple clutches and if not she shouldn't but she needed the capability to exploit the times of plenty. This has a partial similarity to the pragmatic approach above but focussing just on food rather than the totality of 'extremely good habitat' including low predation. Erckmann pointed out that there appeared to be no relationship of polyandry to food supply across the species he surveyed and rejected this hypothesis.

differential parental capacity – this idea was that the female was not really up to looking after the clutch so that the male had to do it. Certainly the male who is capable of looking after his own family finds that doing so is more reliable. There is little evidence that in spotted sandpipers that the female adds benefit after she has laid the eggs (in good habitat). So it is a possibility that the *Actitis* female has gradually lost capability in looking after chicks though unlikely as occasional common sandpiper broods end up with just the female.

Some have suggested that if there is an excess of males that would favour polyandry but in spotted sandpipers there has been no evidence of a routine departure from an even sex ratio. So it is just that in a very good place dominant females grab the resources of prime habitat and prime males. This leaves the poorer (perhaps inexperienced) females to occupy second-rate habitat and hope that a male turns up. As selection theory would suggest the good female succeeds in passing her genes on. Her cost may be that she wears herself out sooner than the common sandpiper female who has the more conventional approach to finding a mate. However, the males of both species end up defending enough space for their brood from other males and staying with their family until the chicks are fledged.

The advantage to the spotted sandpiper male seems to be a slightly lower chance of being cuckolded, a near certainty of eventually getting a clutch, and a bit more choice in the process and less energy expended in fighting.

As the change in tactics has evolved the spotted sandpiper has become slightly smaller than the common sandpiper (all the waders that lay multiple clutches are small), the eggs slightly smaller as a proportion of the female size (as may be reasonable for so many of them to be laid) and the male has become relatively smaller than the female (see Appendix 3 for detail).

Overall we saw in Chapter 1 that the differences in day-to-day behaviour between the two species are quite small. In Chapters 2 to 4 we saw that outside the breeding season the behaviours were essentially indistinguishable other than that they occurred on different continents.

One thing that would not totally surprise me is that in extremely good but patchy habitat a polyandrous population of common sandpipers is found and that on streams at the edge of the spotted sandpiper range their basic behaviour is little different from common sandpipers on streams. Lest it be thought that it is only *Actitis* that show flexibility, during the seminal study of dunlin in Finland the year 1965 had an early spring so the dunlin started breeding early. They successfully hatched chicks early at which point there were three males spare and the early females left their first brood with the male and laid and helped incubate and care for a second clutch (Soikkeli 1967).

As a final comment on flexibility, OLM pointed out that the incidence of polyandry was similar in 1975 and 1976 but for different reasons. In 1975 an adult mink destroyed many nests so there was much re-nesting and this led to females having to lay more eggs and sometimes this was with a series of mates (polyandry) but also some males had their repeat clutch from a new female (polygyny); the average number of mates each male had was 1.43. In 1976 few females turned up (perhaps due to their bad experience in the previous year) and so they were busy laying for all the males that turned up; the number of mates that each male had was then just 1.00. Interestingly in that latter year the average date that males turned up was nine days later than usual and the spread of arrival dates was larger too (up from usual ±10 days to ±15 days) providing a steady supply of males who then also had the highest ever success, with 3.33 hatchlings per male.

One aspect that is theoretically supposed to change between a male versus a female dominant breeding strategy is the sex ratio of chicks that return to their natal area. From 1973–81 OLM reported 30 female to 21 male in spotted sandpipers on LPI. Within our Ashop study the ratio the ratio was 10:10, and including those from reservoirs in the Peak District that joined the Ashop the ratio is 15 males to 14 females, so they appear on that measure to be not very male dominated. This is consistent with all the behaviours we have seen in earlier chapters, suggesting that they could move to polyandry without too much difficulty.

We now need to look at what the habitat is that supports *Actitis*.

THE PHYSICAL HABITAT FOR BREEDING OF COMMON SANDPIPERS

They may breed by streams, rivers, lakes and reservoirs, and near the seashore. The surrounding land uses are wooded, grazed or arable. There is a bare zone between the land and water's edge, usually of a pebbly nature.

They do not breed where there is only a hard edge or the water always flows fast due to the gradient driving it (thus some authors come up with a criterion of a gradient less than around 4 per cent) or where the land–water boundary is totally of emergent vegetation like *Phragmites* or others typical of lowland lakes and slow rivers.

They breed at sea level if a suitable stream enters the sea directly, and at any level where appropriate habitat is found up to 4,000 metres above sea level.

They eat a broad diet of invertebrates gleaned from feeding off vegetation or bare ground or at the water's edge.

The average temperature is around 10°C when they start breeding.

They can use quite a small territory down as low as 100 metres by about 20 metres if it is ideal. But what is the critical physical nature of a territory?

PEBBLES?

The words 'shingle', 'gravel' and 'pebbles' are used fairly indiscriminately in bird literature. Whichever word is used, is the 'pebbly zone' (Figures 1, 10, 50) actually essential? Or is it merely a corollary of erosive action by water? Most areas that we have looked at breeding sandpipers in Britain have a pebbly zone and this has been measured and pondered over. In the Ashop every territory was measured in detail (Yalden 1986b) and later every territory around Ladybower was measured (Yalden unpublished). Over the 40 years on the streams the territories that have been most often occupied and have produced most fledged young have had around 1–2,000 sq. metres of pebbles. But ranking territories from most pebble area to least does not rank from most used to least used, or from most successful to least successful. Observations of time spent feeding (Yalden 1986a and b) showed that although adults appeared to defend a pebbly area they hardly used it but it was used much more by their chicks. But the chicks wanted to be close to safe hiding places, so enormously extra-large areas of pebbles were probably of no extra benefit. The detailed nature of the pebbly area is probably also of relevance: one that is flat and retains moisture in the cracks is probably best. In a survey of 428 km of rivers in Wales it was found that the only factor out of altitude, gradient, rapids, exposed rock, tributary junctions and exposed shingle that correlated with common sandpipers was shingle (Round & Moss 1984).

HIDING PLACES FOR CHICKS?

When the alarm sounds, very small chicks may freeze but the most common response at all ages is to hide by dashing into a maze formed by tree roots, collapsing banks, flotsam or miscellaneous vegetation. A territory must have hiding places. Pebbles and hiding places tend to go hand in hand. During high flows (or significant wind-driven waves) there tends to be enough erosion to dislocate pebbles, and the ground from which they are swept becomes undercut; if there are trees a tangle of roots is exposed, and branches swept downstream become snagged and other flotsam and sprouting vegetation makes places to hide.

NESTING PLACES?

Most nests are among rough vegetation. A closely mown lawn is no good, but the list of places runs from strawberries, tufted grasses, ferns, nettles, maize … – basically whatever provides reasonable cover. In our area the few that have nested direct on the pebbles have been swept away in floods but there could be locations and times where the pebbles do not get flooded and such sites are quoted.

So let us look, with those three aspects in mind, at Figure 54, which shows the centre of the territory which had our most successful male for lifetime reproductive success. In Appendix 1 this is territory 8, which was assessed as being 300 metres long, with a river width of 12.8 metres, 1,944 sq. metres of pebbles, and 1.99 ha of rough grass. It is essentially the same today. In Yalden 1986b another factor was the area of 'improved grassland' to which the farmer applies some treatment and uses for lambs in April. This is well used by the arriving adult sandpipers in April. All the most popular and productive territories in this study area have easy access to such grassland, and it is probably more attractive to adults than the pebbles – but it is never used by chicks; worldwide such grassland cannot be an essential ingredient of a territory. Overall a damp pebble zone does appear to be important in midsummer, when it has lots of emerging small insects which are easy for a chick to catch. This sort of pebbly zone occurs on reservoirs and lochs, particularly where a stream flows into it, and such areas appear to be favoured by chicks.

Assessment of the territory characteristics around the 33 km edge of the reservoirs formed the basis of a project for Rachel Foley, one of Derek Yalden's final year zoology students in 1998. From 1991 to 1997 Derek had scored each territory as occupied or not, and alarmed with chicks or not; he gave a breeding success score potentially running from 0 to 14 over the seven years. He assessed every 200 metres of the reservoir edge for shingle, boulders, undercutting of bank, grass, woodland, and input streams on scales of 1–4. Foley then ground through statistical tests, looking at one variable at a time and

Figure 54: A good territory: one bank is actually a hard-engineered structure, and above that up the bank is the edge of a main road (A57) whose rough vegetation is good for a nest, because it will not be flooded. Few people will go up there as it is difficult to access and noisy with traffic, and ground predators have a good chance of being run over by a car if they access from above. The sitting bird has a good view, and its mate stood on the wall also has a good lookout place. Because the left bank is concrete all erosion occurs on the right bank, which results in many hiding places. The bed of the stream is like a series of small pools with the stream percolating among pebbles. In wet winter or a flash summer flood, the water flow fills the channel to around 15 metres width, but most of the breeding season it is as in the picture; this low flow is mainly due to the fact that 250 metres upstream (behind the photographer) there is a weir, and most summer flow is diverted to an aqueduct feeding directly to the water treatment plant.

at reasonable multivariates, but she found no significant correlations, and few that even looked vaguely important. At the extremes, however, she found that territories with no shingle were hardly ever occupied and that ones with neither grass nor shingle never succeeded. Up the Ashop there is no woodland in any territory, but around the reservoirs there is; it was used by the birds but it was not an attraction.

Overall in the 1998 thesis there were judged to be 82 territories and 20 stretches called pseudo-territories which were big enough lengths of bank but were never occupied. (Typically, 50 of the territories were used in any year from 1991 to 1997 inclusive; the minimum in those seven years was 44 and the maximum 54.) Interestingly when the population went to a very high level a decade later (Figure 44) it peaked at 84 in 2007 and 79 in 2011 before returning to usual levels. It was reasonable to claim that the area was saturated at 84, with every territory used. There were better territories up the Ashop that were vacant, and these have been re-colonised by common sandpipers since 2010.

A very interesting study of habitat was reported from near Salamanca in Spain (Diez & Peris 2001). They sampled 88.9 km of tributaries of the Duero and Tagus rivers at around 1,000 metres, and they had sandpipers all the year. In the breeding season they found the main correlation was with the width of the river and its edges (shingle), and this was also generally true in winter – but there then also appeared a correlation with the presence of livestock, and they suggested that their organic waste provided the habitat for winter insects. Presence of sandpipers was independent of water chemistry, and this appears to be generally true. In France a correlation with the presence of grayling (*Thymallus thymallus*) was reported (Roche 1989), but in England most rivers with grayling have no sandpipers, and the sandpiper streams have no grayling. This again is indicative that it is the physical structure of the water/land boundary that is important (those largish rivers in France had braided structure; grayling like a pebbly bottom and flowing water, so there is clearly a potential for correlation between the species) and that water chemistry is unimportant.

Height above sea level appears to be irrelevant. Both species will nest near the sea shore; BNA reports spotted sandpipers up to 4,700 metres and BWP reports common sandpipers to 4,000 metres.

However good the physical habitat is, they will not prosper if there is too much disturbance from predators and competitors (Chapter 6). This excludes them from suburbia with its cats, dogs and rats and busy recreation areas.

Sandpipers are somewhat attracted to join a flourishing population (Chapter 7) so a territory that is isolated from others is not as good as a group of territories giving a degree of communal support.

The importance of active creation of habitat by weather and water was nicely illustrated by a paper describing the changes in some tributaries of the Vistula in south Poland, where a detailed survey was done in 2007–2009 comparing 'natural' sections of the river with 'regulated' sections; at that time there were 10 pairs of sandpipers on the former and only 3 on the latter. A massive flood occurred in 2010, returning the whole river to 'natural', and in 2011 there were 16 and 10 pairs respectively (Kajtoch & Figarski 2013).

WHERE MAY THE BREEDING BEHAVIOURS DESCRIBED BE OBSERVED?

BWP on the common sandpiper – 'Prefers fresh water of clear lakes, rivers, or streams, especially favouring fairly fast flowing rocky upper courses of streams and rivers, with fall of up to approximately 40 metres per kilometre and subject to sudden spates or flooding alternating with periods of low flow.'

BNA on spotted sandpiper – 'Occupy almost all habitats near water, everything from the shorelines of wild rivers and lakes to urban and agricultural ponds and pools.'

Most birdwatchers know where they would expect to find sandpipers breeding in their home country, and as seen in the standard textbook descriptions above they are very adaptable; basically they like waterside habitat that is renewed each year by vigorous water erosion or that is newly created in some other way.

UNUSUAL BUT TEMPORARY SITES

Most species are always on the lookout for new places to try breeding. Thus sandpipers have been found in many places where the general local opinion was that it is 'not suitable habitat', mainly because they have not been there before.

So let me describe some 'not suitable' places that I have visited, where common sandpipers bred during the 40 years covered in this book (my visits were not usually coincident with the breeding events):

a) One of our colour-ringed birds was reported breeding two years running at a big landscaping project to tidy up the spoil tips from coal-mining activity at Annesley (53°05'N; 1°16'W) in Nottinghamshire. This work had produced 'a lake' with 'shores of coarse spoil', and the work in progress was fenced, so human disturbance was minimal. An adult that had previously been ringed in the delectable surroundings of the Peak District decided to give it a go.

b) Where a little river, the Conder, about 3 metres across, enters the Lune estuary salt marsh (where I ringed passage birds from 1992 to 2005, as described in Chapter 2) on a pebbly bend, and where the water was still normally fresh, a pair hatched chicks quite unexpectedly in one year. I do not think the chicks fledged, but there were mute swans on that bend, so the sandpipers may have moved well away.

c) A major flood alleviation scheme was done in the Mersey valley running through south Manchester, which involved digging large lagoons which later became water parks, which are now big recreational areas. When new, Sale Water Park had a fenced-off area with shingle, and in the 1980s common sandpipers turned up and bred.

d) A pair was heard alarming near Horsham in Sussex in 1978 and chicks were seen; over the next few weeks alarms persisted, though getting views of chicks proved difficult. This was on the River Arun, which at this point had meanders and was clearly suitable that year for breeding (Sussex County Records, Sussex Ornithological Society archive, Chichester).

e) On St Agnes, one of the Isles of Scilly in the Atlantic Ocean off the far south-west of the British Isles, chicks were raised to fledging in 2008 (Isles of Scilly Bird and Natural History Review). This is a small island, about 50 ha in extent, with little to recommend it as sandpiper habitat – no river or lake, a tiny trickle of a 'stream' – and the birds often fed on beaches by the ocean.

These observations are consistent with other colonisations of new habitat.

f) In the Netherlands, some newly landscaped area attracted them (Erhart 1997).

g) In Scotland, when open-cast mining operation required the diversion of a river and its subsequent re-diversion a few years later as part of landscape restoration, the birds rapidly re-colonised to a high density (Bray et al 2012).

h) The most amazing records are of breeding in Africa. The best attested are a nest found by Lake Victoria in August, and from chicks seen in Kenya in June (Benson & Irwin 1974). On 30 April 1959 a pair shot in the Congo showed 'a male with growing testes and the female with active ovaries' (Curry-Lindahl 1960).

i) The Rev. Gilbert White reported breeding near Selborne in 1768. This seems an unlikely place, but about a mile north of his church, where the small river meanders, there are some lakes that look as though they were around in the 18th century and the combination of stream and lake would have been usable.

Table 9.1 below summarises all the records over the last century for common sandpipers from county avifaunas for parts of England generally thought 'not suitable'.

COMMON SANDPIPER BREEDING RECORDS IN SOUTHERN AND MIDLAND ENGLAND

Table 9.1

County	Years bred	Young fledged	Source
Bedfordshire	1904, 1970	unknown	Trodd & Kramer 1991
Berkshire	1934, 1949–1957, 1991–1992, 1994–1995	yes: in 1934, 1953, 1994 and 1995	Standley et al 1996
Buckinghamshire	1929	unknown	Ferguson 2012
Cornwall	last bred 1910 (until 2008 on St Agnes – see text)	probable	Penhallurick 1969
Devon	regular to 1962, irregular since	-	Sitters 1988
Essex	1947	unknown	Wood 2007
Gloucestershire	1993, 1999, 2001	probable	Kirk & Phillips 2013
Hampshire	1978	no	Clark & Eyre 1993
Hertfordshire	1890, 1910, 1954–57, 1958. 1961, 1967, 1969	probable	Sage 1959, Gladwin & Sage 1986
Leicestershire and Rutland	1885, 1906, 1908, 1998	unknown	Fray et al 2009
'London' area	1950–58 (mainly duplicates of Hertfordshire records)	yes; in 1954,1955, 1957	Montier 1977
Norfolk	1897, 1912, 1928, 1962, 1963,1980	probable	Taylor et al 1999

Table 9.1 (*cont.*)

County	Years bred	Young fledged	Source
Oxfordshire	on Thames before 1947. 1985 gravel pit	probable	Brucker et al 1992
Somerset	regular until 1920, last 1970	yes	Ballance 2006
Suffolk	1899	chicks seen	Ticehurst 1932
Surrey	1895–1896, 1911, 1969–1970, 1981 and 1987	unknown	Wheatley 2007
Sussex	1978	probable (see text)	James 1996 Sussex Records Chichester
Warwickshire	2002	unknown	Harrison & Harrison 2005
Wiltshire	1980	unknown	Ferguson-Lees et al 2007

SPOTTED SANDPIPERS IN SCOTLAND

A pair bred in Scotland in 1975, where a freshwater stream entered the sea. There was a pair of common sandpipers nearby with chicks; the pair of spotted sandpipers was seen and a nest of four eggs found on 15 June that were incubated until 2 July, but it was then deserted, possibly due to a combination of heavy rain and cattle. The eggs were sent to the museum; two were developing, but two were infertile (Wilson 1976).

MIXED SPOTTED AND COMMON

A male spotted sandpiper was seen copulating with a common sandpiper at a gravel pit at Elland in West Yorkshire on 5 June 1990. It displayed with wing-raising and walking around the female followed by mating, after which it flew off calling, and the female remained splashing and preening at the water's edge (Lawrence 1993).

Between 30 June and 7 July 1991 a common sandpiper with three chicks that fledged on 4 July was accompanied by a spotted sandpiper at Southern Washlands nature reserve near Wakefield in West Yorkshire (Ogilvie 1994). This site is only 30 km from the gravel pit at Elland; both are by the River Calder, and both are unusual sites. Whether the spotted was the father or just keeping company is not known.

On north Mainland Shetland on 30 June 1998, what appeared to be two pairs of common sandpipers turned out in one case to be a mixed pair with the spotted alarming from a rock; it was relocated five days later nearby (Shetland Bird Report for 1998).

WHERE DON'T THEY BREED?

This is just a brief comment on areas where the habitat looks superficially plausible (to me) but there is documented absence. Kruckenburg et al 2012 listed the 15 species of wader breeding on Kolguev Island (69°N) in the Barents Sea. The paper includes pictures of a river where Temminck's stint (*Calidris Temminckii*) and common ringed plover (*Charadrius hiaticula*) breed; these places look suitable for common sandpipers, but they are not there. Iceland (65°N) has none, nor do the Faroe Islands (62°N). Sandpipers do breed at 70°N in Norway and the Kola peninsula, and at 72°N in Siberia (Lappo et al 2012) so latitude *per se* is not an issue, though once the Arctic tundra is reached, this is not used.

I believe that crossing significant distances over oceans on their northwards spring migration (see Chapter 4) is difficult for them, and anyway the populations in their core habitat are not overcrowded enough to force them to move further on. Indeed at the present time there are unused territories in the British Isles and on mainland Europe, so that pushing on towards Iceland or on to islands north of Russia would be daft. However, as the world warms and the busy parts of the world get busier, there are new places for them to colonise.

HUMAN DISTURBANCE

The study areas in the Peak District are well looked after by the Peak Park authorities, the farmers and the water companies. Day trippers are led to car parking areas and picnic sites that are not sensitive. Footpaths are mostly far enough away from the habitat used by sandpipers. Some extremely popular areas in the Peak, like Dovedale, may have had sandpipers in the distant past and may have them again if humans disappeared – but as implied in earlier chapters, unravelling all the factors affecting them is difficult and we need to be thankful that the reservoirs and streams are still as good as they are. Day trippers are mainly around from 10.00 to 16.00 hours, so as long the birds can find refuges they will usually survive the day trippers; but once housing estates and holiday cabins encroach and dog walkers are out from dawn to dusk, there is little hope for sandpipers.

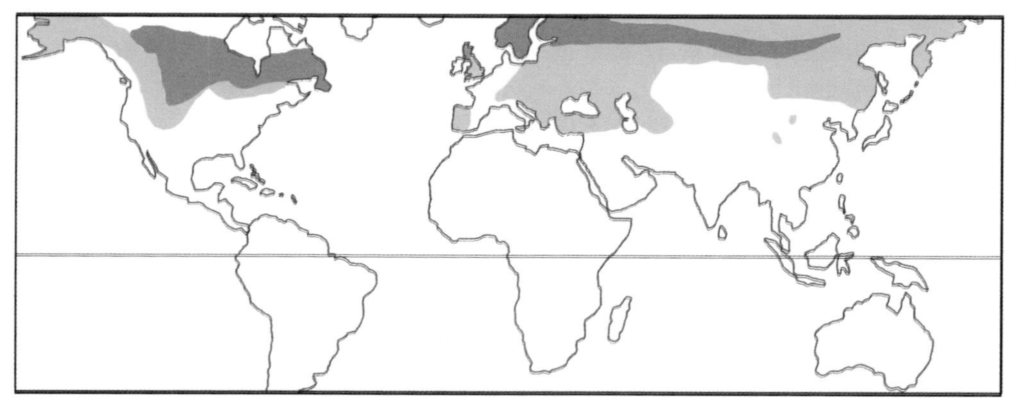

Figure 55: A broadbrush picture of their breeding distribution estimating their abundance at two levels - abundant and quite abundant.

SPOTTED SANDPIPER PHYSICAL HABITAT

The words used to describe spotted sandpiper habitat in BNA are 'everything from the shorelines of wild rivers and lakes to urban and agricultural ponds and pools'. In more detail, the three characteristics of good habitat are specified as 'shoreline for foraging … semi-open habitat for nesting … dense vegetation for brood cover', but sandpipers may 'nest away from shore, feed in undefended areas, and secure water's edge after the chicks hatch'. So a specific area set aside for the chicks is most crucial, and explains why when the density gets extremely high, as it did on LPI, chicks get killed by other spotted sandpiper adults. What would be really nice is a study on 'urban ponds and pools', as this is not a category adopted by common sandpipers for breeding.

It should finally be reiterated that the detailed studies of common sandpipers have been undertaken mainly on streams, while those of spotted sandpipers have been on islands. That should probably not be taken as representing the requirements of the birds, but more a matter of human convenience for study areas. Nevertheless all the high densities of spotted sandpipers are reported from low predator islands – LPI (around 10 per hectare), Great Gull Island (around 1 per hectare, but sharing with thousands of terns) and Belle Isle in the River Detroit (around 6 per hectare, Miller & Miller 1948).

FINALLY

In the 1990s we found it difficult to explain how common sandpipers sustained themselves in the Peak District, as their success at fledging seemed inadequate to replace their losses of adults. Slowly it dawned on us that they were not quite as site faithful as we had assumed; and then, as we got birds returning that were 15 years old and evidence accumulated of movement of successful adults to elsewhere, we were happier with a high estimate of adult survival. We always knew that it was difficult to see young sandpipers, and thus our fledging estimate was probably on the low side. But gradually we have also come to the view that English streams are second-class habitat. Indeed the reasons that these sites are good for bird ringers to catch adult birds and find chicks are probably the reasons that other predators find them good, too.

Nevertheless it is clear that these sandpipers do not produce a big surplus of new recruits anywhere that would lead to rapid population growth. Indeed this seems frequently the case for shorebirds, and is probably the reason why so many are declining. The scarcity of common sandpipers on the far islands of Scotland and their absence from the Faeroes and Iceland suggests that they are not expanding their range now – and indeed they are retreating from Ireland.

It is reasonable that on a world scale they are of least concern, as they still number in millions and there is only evidence for local declines. It is reasonable that in the British Isles they are amber listed, as they are declining. But the argument presented in Chapter 8 is that they are possibly retreating from their maximum ever. If more people need housing and a priority of the people is protecting sandpiper predators, then some further reduction is likely. But it is likely, too, there will always be enough common sandpipers to give pleasure, as it is the most widespread sandpiper in the world and indeed the only one likely to be seen by non-birders while out for a walk or picnic in the summer.

In North America, the less densely populated country would appear to make the spotted sandpiper fairly secure.

For both species the loss of wintering habitat in mangroves and ever-increasing human populations may be a concern, but most countries have become more aware of the benefits of mangroves for protection of the coast and for fisheries' productivity. The detail of how sandpipers use mangroves would be a valuable study to do. On migration and also in other wintering places, they appear to use such diverse and widespread habitats that the threat to them seems small.

APPENDIX 1: DESCRIPTIONS OF BREEDING STUDY AREAS

Each of these studies has birds with coloured bands/rings so they could be followed as individuals.

Units – the source references use different units. All numbers in the following have been written as metric (using 1 metre = 3.3 feet; 1 km = 0.62 miles, 1 ha = 2.5 acres).

Little Pelican Island (LPI), Leech Lake, Minnesota, USA: 1972–1990

The Minnesota landscape is a rich mosaic of lakes, forest and farmland and is sometimes called the Land of 10,000 Lakes (the US Department of Natural Resources says there are 11,482 lakes larger than four hectares in Minnesota). Leech Lake is one of the largest, with around 300 km of shoreline. Mostly the land around it is about 400 metres above sea level, with hills up to about 580 metres, and has very old hard rocks overlain with deposits from the ice ages. The lakes are covered with ice in winter and the spring thaw leads to ice-free conditions around the end of April. It has a 'continental' climate with average July temperatures of 20°C, and dry. Leech Lake drains into the Mississippi 2,000 km upstream of its delta.

This is in an area of logging, tourism and some farming. The largest city nearby is Bemidji with 13,500 inhabitants. Leech Lake has resorts with a focus on fishing supported by stocking with millions of fry (mostly Walleye). There is a large dam at the north-east end of the lake built in 1882 which raised the level by up to 2.2 metres and merged several lakes into one: this is water supply for Minnesota. The railway came in 1896. The last Indian skirmish was the Battle of Sugar Point in 1898, which the local tourist information wryly points out is little mentioned probably because the Indians won. The lake is now part of the Ojibwe Indian Reservation. It was untouched by outsiders until fur-trappers and traders arrived in 1785 – these first Europeans called it Bloodsucker Lake but it is now Leech Lake. People are advised how many fish they can eat before mercury becomes an issue (the mercury comes from distant burning of fossil fuels whose emissions are deposited on trees and washed off by rain and snow) but essentially the lake is pure. During the breeding season there are many boats of various kinds using the lake.

There are a few islands in the lake. M&O give the key facts for LPI as 47°07'N, 94°22'W, 1.6 hectare area, with a shoreline of about 650 metres. It is 7–8 km from shore. The larger Pelican Island is 36 hectares and only 200 metres away, and had a few spotted sandpipers breeding; Gull Island, 0.3 hectare, is 300 metres away had no sandpipers. Figure 56 shows a photograph of LPI in the 1970s. The island now lacks vegetation – see Google Earth – and lacks sandpipers. This island contained around 20 nests during the peak years, and the remaining islands and lake shore, totalling over 300 km, only held a few more. LPI's vegetation is described as semi-open herbaceous, dense grasses and sedges, cattails, shrubs and trees. The biologists had four purpose-built 3-metre observation towers so the birds could be watched with minimal disturbance, and the island was almost a private outdoor laboratory. Predator control was done sometimes; especially removal of mammals. The spotted sandpipers were caught in shoreline funnel traps or mist nets. The work was funded by a series of grants to professionals and biology students under Professor Lewis Oring of the University of North Dakota. In the text this study area is usually referred to by the abbreviation LPI.

From 1982 the area studied was expanded so that more could be learned about where birds born on LPI might settle and where adults that moved might move to. The study then included (Big) Pelican Island, and two stretches of beach on the mainland: Whipholt and Timberlane.

Peak District, England, 1977–2016

The West Country excepted, the English landscape to the south of the Peak District has no hills higher than 300 metres above sea level, and even those hills are rare. Mostly this southerly part is below 80 metres: a mixture of farms and towns and cities with rivers that are slow-moving usually with hard flood banks, and at times of exceptionally high rainfall they flood over into the surrounding land rather than eroding the banks. The soils are largely ones left after the ice ages. The Peak District is the first 'upland' as one goes north. It still only reaches 600 metres at its highest, but this leads to higher rainfall and the steeper slopes on mixed rocks of millstone and shale lead to faster flows and streams with changing meanders and pebble/shingle areas. The tops of the highest hills are covered with peat and tundra vegetation occupied by golden plover and dunlin. There is rarely ice on the streams or reservoirs, and though serious snow can occur in winter the area is generally green throughout April. As in Minnesota, the warmest month is July but the average temperature is only 14°C and from May to July it tends to rain for some time every other day.

The Peak District is close to major cities (Manchester, Sheffield, Leeds, Stoke-on-Trent, Derby) and accessible to upward of 10 million people on a day trip. Every valley has been dammed to make reservoirs, and there are major roads crossing the Peak District which are busy with business people and tourists. The valleys are farmed with sheep and cattle and the moors are managed for red grouse (*Lagopus lagopus scoticus*) and some of the valleys are commercial forestry areas. Crows, foxes and mustelids are controlled by gamekeepers and farmers. There are footpaths, cycle tracks and bridleways for horse-riders. But the Ashop Valley, though a busy road, has few places to park motor vehicles, and the farmland through which the streams flow does not have public access; furthermore, because of the presence of sheep, dog-walkers are rare and keep their dogs on leads. The land is mostly owned by the National Trust, who look after it responsibly

Figure 56: Stephen Maxson kindly allowed me to re-use his photograph of LPI from Maxson & Oring 1980.

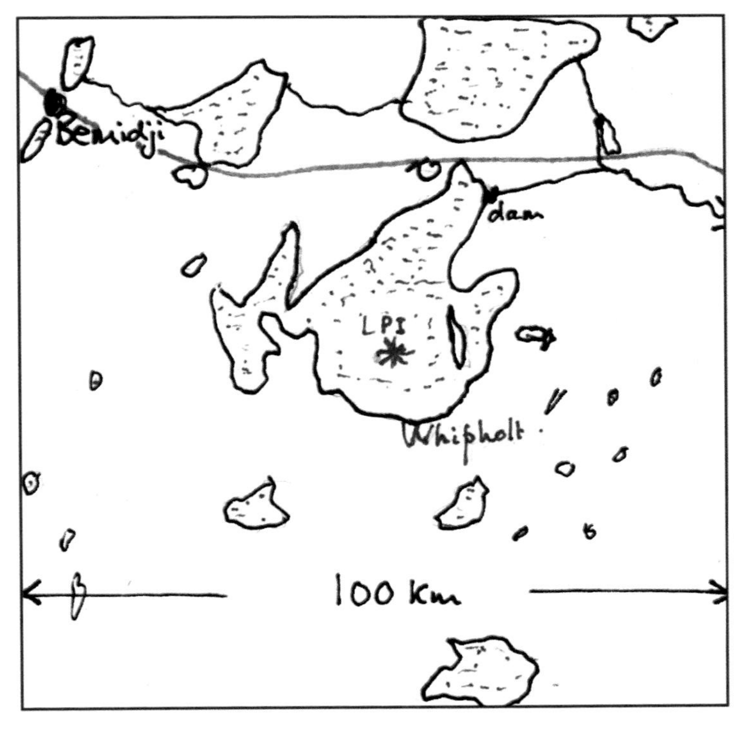

Figure 57: Sketch map of Leech Lake and environs.

Figure 58: This general view of the River Ashop is looking downstream, with territory number 5 on the shingle below the actively eroding bank immediately in front and territory 4 in the distance.

with their dedicated farming community. There is no industry other than quarries in the limestone areas to the south, but in the industrial revolution air pollution from the major industrial cities led to deposits on the hill tops particularly of sulphur, so streams that are already naturally acidic from peat have had an additional load of sulphur; fortunately this has been much reduced in the last 20 years by flue-gas desulphurisation being fitted to coal-power stations and by the closure of old plants. Farming is sheep and cattle grazing and does not appear to be an issue. The faster stretches of the streams have dippers, which are more sensitive to stream pollution than sandpipers. In the 1970s the area was almost without birds of prey because pesticides and keepers had removed all peregrines, buzzards and merlins, and most sparrowhawks. Some kestrels and recent escapes of goshawks were there, and during the last 30 years birds of prey have returned.

The principal study population centres on where the River Alport (actually a stream only a few metres wide) joins the Ashop (a slightly wider stream but less than 20 metres wide) at 53°23'N. 2°47'W. In the text this is called the Ashop study area. The Ashop flows down to Ladybower reservoir which flows into the River Derwent, which joins the Trent, which flows into the North Sea via the Humber Estuary.

Photograph 58 shows typical stream habitat. The valley is typically less than 100 metres wide and the stream may first be up against one bank where it erodes a steep cliff, and may then meander across to the other side where it erodes again. In territories 8–10 it is very

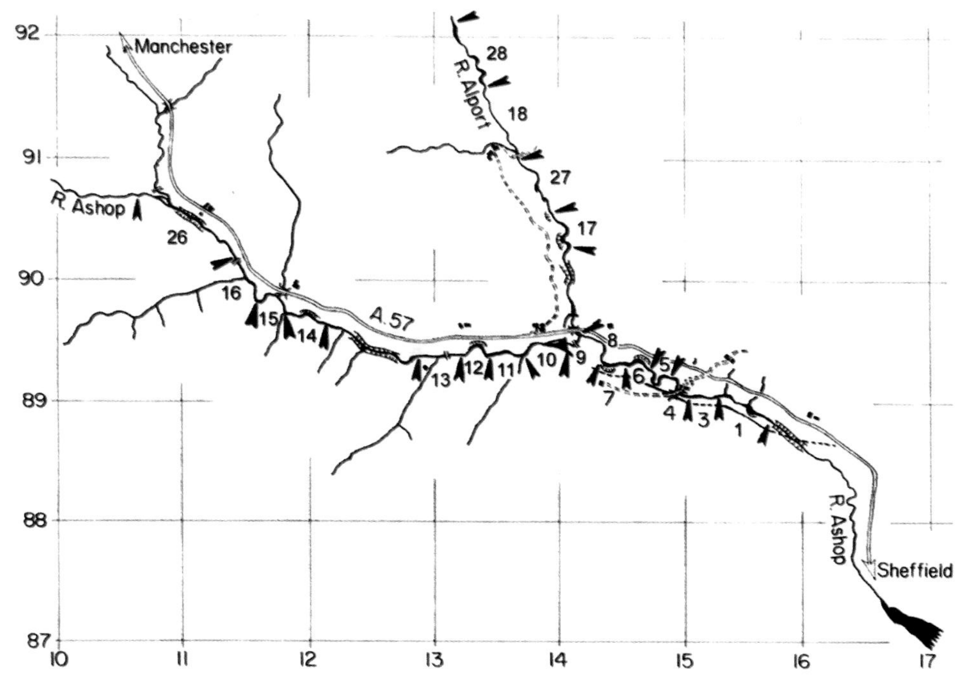

Figure 59: The distribution of territories along the streams has been essentially the same throughout the period though some are unoccupied each year. (The numbering shown is a that used in Holland & Yalden 2012, which was only different from that used in Yalden 1986b to allow for territories used during the 40 years, but not used in the years 1984 and 1985, which were the subject of the 1986 paper. This rather peculiar system originated with an attempt to see the absolute maximum number of pairs we thought could fit in the valley, but this level of occupation was never achieved).

constrained by engineering features – a weir that abstracts water to the aqueduct which takes water direct to water treatment plant, and hard features that stop the A57 road being undercut (Figure 54 in Chapter 9 shows territory 8). Fields of improved grass where the springtime lambs start their life are used by early birds. As the nest sites are spread over 10 km of stream all observations are done on foot, finding convenient bits of bank to observe from. Birds rarely stand still in a place where their legs are visible.

The most productive territories have been t7 (30 fledged); t10 (29 fledged); t8 (26 fledged); t5 (24 fledged); t16 (25 fledged); t14 (24 fledged).

In 2009 when the population sank to its low point of 3 pairs the occupied territories were t5, t9 (historically 20 fledged) and t14. Over the following few years the take-up of territories was t8 and t16 in 2010, then t9, t10 and t17 in 2011, then t6, t7 and t11 in 2012. The take-up of t17 was continued through to 2016, which is a good sign of return to health of the population, since the Alport is a very narrow stream for common sandpipers. The most productive territories historically were generally the first to be taken up again. The stream moves its path a little every time there is a big flow, but this only revises the detailed position of shingle and not the general situation since around 1970 (other river courses that were used earlier than that are discernible by damp flushes).

About 3 km downstream of territory 1 in Figure 59 the Ashop enters the west arm of the Ladybower reservoir. The main volume of the reservoir is up the Derwent valley, and higher up there are two more reservoirs: Derwent and Howden. These reservoirs were a subsidiary study area with 33 km of shoreline. In the text these three reservoirs are usually abbreviated as **LDH**.

Other reservoirs and streams have also been sampled, and interchanges of birds with them are shown in the text at Figures 42 and 43. Sometimes in the text **the Peak District** is referred to when discussing information gathered throughout that area.

The reservoirs were formed by sequential completion of dams at Howden in 1912, Derwent in 1916 and Ladybower in 1945. These removed most people from the upper Derwent, with new housing being constructed down the valley below Ladybower Dam. But during construction of the reservoirs a significant temporary town near Derwent, coupled with the changing nature of the valley, may well have made the area quite unattractive to sandpipers for 30 years. The area has a massive influx of tourists though they tend to be in two extremes: keen walkers who disappear to the hills, and sedentary folk enjoying the views. The edges of the reservoirs are difficult of access in most places and not encouraged by the water company. Fishing is only allowed on Ladybower, and no boating or swimming is allowed anywhere. A general view of a reservoir is shown in Figure 60.

The work was done by amateurs in a ringing group which fortunately included Derek Yalden, the Manchester University zoologist who modestly described himself as amateur,

Figure 60: Photograph of Howden reservoir; the highest in the Derwent Valley (LDH)

since his specialism and what he was paid for, was mammals and general zoology; sandpipers were his hobby.

Moorfoot Hills, Scotland (1993 to the present)

Tom Dougall from Edinburgh has done a similar study, and during two years was joined by Allan Mee doing a PhD at Sheffield University. The Moorfoots, just south of Edinburgh, are typical of a large part of southern Scotland where hills are rounded and less than 700 metres in height. The land use is fairly similar to that of the Peak District, but although Edinburgh is close the total human pressure is much less. The study population breeds along streams at around 55°45'N, 3°01'W that look very similar to those in the Peak District. Indeed all the streams draining the hilly country running from Edinburgh to Manchester, 300 km away, have quite similar populations of sandpipers. Most of the scientific papers from 1995 onwards have been joint (e.g. Dougall et al 2010).

Another PhD on stream waders was done by Shirley Jones from 1979 to 1981 at Durham University, working on the Upper Tees about halfway between the Moorfoots and the Peak District; nothing she reported about common sandpipers was inconsistent with the long-term studies. Many of her observations were used in BWP.

Key Aspects of the Study Areas

Table A1.1

LPI	Peak District	Moorfoot Hills
Quite central in breeding range, though towards southern limit	On very edge of a declining British breeding range	Close to edge of world range, but more central within British Isles
An island, and only about 20,000 resident people in 50 km radius (though significantly increased by tourists in summer)	Major recreational and farming area with several million residents in 50 km radius	Quite secluded but only 40 km from Edinburgh (about 1 million residents in 50 km radius)
1.6 ha island in large lake	10 km of stream through grazed land	15 km of stream through grazed land
Professional full-time	Hobby, part-time	Hobby, part-time
Targeted predator control	Incidental predator control	Incidental predator control
Nearly all spotted sandpiper in the area were on this island at a very high density, and when the island was invaded by bigger birds the sandpipers left.	The surrounding reservoirs have plenty of sandpipers, and some other streams and rivers too, at a fairly average density	The surrounding hill streams and rivers are all sandpiper areas

STATISTICS

This book has no reference to statistical tests, as it is written for people who are interested in bird behaviour. Within the cited references on *Actitis*, readers who want statistics will find use of Fishers exact test, Fisher sign test, G-test, Mann-Whitney U, Meddis, paired t-test, Combined Cohort, Pearsons, Rice's conditional binomial exact test, Spearman rank correlation, stepwise regressions, QAIC, Wilcoxon signed-rank test, Z-test.

ARCHIVE

Logs of visits to the Peak District and other data have been archived with the BTO.

APPENDIX 2: A LIST OF FOOD ITEMS

This is a list of items detected in the diet of common sandpiper (CS) or spotted sandpiper (SS) to whatever level they are identified. The How column specifies direct observation (Obs), deduction from observation (R.Obs), stomach of dead bird (Stm), Faeces (Fc) or Pellet (Plt). In the Habitat column B = breeding, thus freshwater and terrestrial, P = passage, thus various habitats, W = winter, mainly brackish/marine. If you would like more detail please find the appropriate reference. Absence of an item from this list does not imply it is not eaten. This is not an exhaustive compendium; BWP, BNA and Glutz 1977 give good lists from historic sources, and the following tabulation is mainly from more recent work: Arcas (Arcas 2000, Arcas 2001, Arcas 2004), DY86 (Yalden 1986a), M&O (Maxson & Oring 1980), PKH (my casual observations) and a few others as given.

Table A2.1

Item	How	Where	Habitat	**When**	**Ref**	**SS or CS**
WORMS						
Nereis diversicolor	Obs, Fc, Plt	Spain, Sussex, Tidal river	P and W		Arcas, PKH	CS
Lumbricidae	Obs, Stm, Fc	England Fields by river	B	April–May and if wet	DY86, PKH	CS
MOLLUSCS						
Hydrobia Ulva	Plt	Spain Tidal river	P	Autumn	Arcas	CS
Gastropoda	Fc	England freshwater	B		DY86	CS
CRUSTACEANS						
Amphipods	R.Obs	Minnesota Freshwater lake	B	Spring and Summer	M&O	SS

Item	How	Where	Habitat	**When**	**Ref**	**SS or CS**
	Stm	Porto Rico	P and W	All year except June	Wetmore 1916	SS
Orchestia gammarellus	Fc, Plt	Spain	P		Arcas	CS
Sphaeroma	Fc	"	W		Arcas	CS
Carcinus maenas	Fc, Obs	" Sussex	W		Arcas PKH	CS
Talitrus saltator	Fc	"	W		Arcas	CS
Ciathura carinata	Fc	"	W		Arcas	CS
Uca	Stm Obs	Porto Rico Guinea-Bissau, Algarve	P and W	All year except June	Wetmore 1916 Zwarts, PKH	SS CS
Corophium volutator	R.Obs	Britain	P	July–August	Holland 2009	CS
Corophium multisetosum	Fc	Spain	W	Nov–Feb	Arcas	CS
Ligia italica	Obs	Malta	P		Sultana & Gauci 1982	CS
Ligia oceanica	Fc	Spain	W	Oct–Feb	Arcas	CS
ARACHNIDS						
Aranae	Fc	England	B		DY86	CS
Opiliones	Fc	England	B		DY86	
" nymph	Fc	England	B		DY86	CS
Misc spiders	Obs	England River bank	B, P and W	all year	PKH	CS
INSECTS						
Ephemeridae	R.Obs	Minnesota	B	Spring and Summer	M&O	SS
Heptageniidae	R.Obs	Minnesota	B	Spring and Summer	M&O	SS
Plecoptera Adult and nymph	Fc	England	B		DY86	CS

Item	How	Where	Habitat	**When**	**Ref**	**SS or CS**
Orthoptera *Scapteriscus didactylus*	Stm	Porto Rico	P and W		Wetmore 1916	SS
Hemiptera – Corixa	Stm	Porto Rico	P and W	All year except June	Wetmore 1916	SS
– Mesovelia	"				"	
– Aphididae	Fc	England	B		DY86	CS
Coleoptera larvae	Fc	Spain	P		Arcas	CS
Cicindelidae	Stm	Porto Rico	P and W	All year	Wetmore 1916	SS
Carabidae –ad and larvae	Fc	England	B	April–June	DY86	CS
Dytiscidae	Plt	Spain	P		Arcas	CS
Agabus and Hydroporus	Plt	England	B	July	Smith	CS
Hydrophilidae	Stm	Porto Rico	P and W	All year except June	Wetmore 1916	SS
Berosus	"				"	SS
Staphylinidae (rove beetles)	Plt	Spain	P		Arcas	CS
Omaliinae	Fc and Plt	England	B		DY86 Smith	CS
Byrrhidae						
Byrrhus pilula	Pt	England	B	July	Smith	CS
Elateridae ad and larvae	Fc	England	B		DY86	CS
Cantharidae	Fc	England	B		DY86	CS
Curculionidae –ad	Fc	England	B		DY86	CS
Diptera *Paracoenia turbida*	Obs, Fc	Yellowstone	P		Kuenzel and Wiegert 1973	SS
adults	Fc	England	B		DY86	CS

Item	How	Where	Habitat	**When**	**Ref**	**SS or CS**
Tipulidae	Stm	R,Ussuri USSR	B	Spring	BWP	CS
"	Plt	Spain	P		Arcas	CS
larvae	Fc	England	B		DY86	CS
Chironomidae	R.Obs	Minnesota	B	Spring and Summer	M&O	SS
"	Fc	Spain	P		Arcas	CS
larvae	Fc	England	B		DY86	CS
Lepidoptera larvae	Obs, Fc	England	B		DY86 PKH	CS
Adult moth	Obs	England	B		DY log	CS
Trichoptera Adult and larvae (caddis)	Fc	England	B		DY86	
Hymenoptera adults	Fc	England	B		DY86	CS
Proctotrupoidea	Fc	England	B		DY86	CS
Formicidae	Fc	England	B		DY86	CS
Myrmicinae	Plt	Spain	P		Arcas	CS
VERTEBRATES						
Hemidactylis bowringii	Obs Gecko	China	W	January	Stanton 2013	CS
Fish	Obs	Switzerland, China,	P		See text	CS

APPENDIX 3: BIOMETRICS

There are several quoted datasets of biometrics, and they are all similar but slightly different. Their relevance is to the questions

How different in size are the common sandpiper CS and spotted sandpiper SS?

How different are males and females?

What size is the egg relative to the bird?

The following have biometrics which have been widely used: Glutz 1977, Prater et al 1977, Lofaldi 1981, BWP, BNA.

LINEAR DIMENSIONS

Wing-length is measured from the carpal joint to the tip of the longest primary with the wing folded and the primaries straight (a standard ringing technique). Some measurements are on skins, some on recently killed birds and others on live birds.

For male CS the means quoted are:

110.4 Glutz skins from Siberia

111.5 Prater skins from BMNH

108.7 Lofoldi breeding birds in Norway

112 BWP

Lofoldi had a small sample (13) which included a bird at only 102 mm, which is much smaller than any we have ever had in 40 years. Our mean over 40 years is 110.5. All these sets have a standard deviation around 3; anyone who claims they are measuring to better than 1 mm in the field is delusional, so the sensible average length is **111 mm**.

For male SS the means quoted are:

105.6 Prater

105 BWP

BNA says 'No comparative studies available', so a sensible assumption is **105 mm**.

For female CS the means quoted are:

113.7 Glutz

112.5 Prater

115.3 Lofoldi

112 BWP (it is odd that this is the same as the male, since all other sources have the female bigger)

Our mean has been 115.5, so a sensible assumption is **115 mm**.

For female SS the means quoted are:

109 in both Prater and BWP so **109 mm** is sensible.

Bill-length is measured from the last feather by the upper mandible to the tip. All quoted averages for CS (except Lofoldi for his 13 males, 23.4, and again a minimum of 20.9 – much smaller than anyone else; perhaps it was a stray Temminck's stint) – are between 24.5 and 24.9; it is useless for discriminating sex. We rarely bother to measure it as we want the bird back on its territory, but a mean of about **25 mm** is sensible.

For SS Prater quotes 24.1 mm for both sexes. BWP says 24.1 for females and 23.2 for males so **24 mm** is sensible.

Tail-length is not nice to measure on a living bird. Prater quotes 55.5/55.6 for CS female/male and 51.1/49.2 for SS, with BWP quoting 53.5/52.5 versus 48.7/46.1. Any twitcher will tell you that SS have shorter tails, so **5 mm** shorter seems reasonable.

Tarsus

For CS, Prater gives 23.6 regardless of sex, BWP gives 24.5 for females and 24.3 for males so **24.2 mm** is reasonable. The few that we have measured are similar.

For SS, Prater gives 23.5 regardles of sex, BWP gives 24.1 for females and 23.5 for males so **23.8 mm** is reasonable

Eggs

CS eggs are quoted in BWP as **36 mm** long and **26 mm** in diameter.

Glutz gave more information:

Norway 35.21 × 26.03

Sweden 36.28 × 26.17

UK 36.40 × 26.67

Japan 36.92 × 22.55

The few we have measured do not contradict the European ones.

SS eggs are quoted in BNA as **32 mm** long and **24 mm** in diameter, and this came from a detailed study of 548 eggs by Reed & Oring 1997 on LPI. BWP quotes 32 × 23 from museum collections.

Mass

Male CS is quoted in BWP as 45.5g on the breeding ground (and 40.5g in winter). We have weighed 156 in the Ashop; they give a mean of 50.2g and only 16 were below 46g. Lofoldi did not separate male and female, but his histogram of all masses only has four birds below 46g and is consistent with **50g.**

Most female CS are caught with eggs inside them, so I have only used birds in April and June (41 birds) to deduce a 'normal' breeding season mass of **57g** (with eggs, their mass can be anything up to 92g). BWP quotes females at 4.5g heavier than males.

The standard deviations are around 3.5g.

For male SS, BNA gives values for each stage in the breeding season, with a minimum of 40.1g the day they arrive and a maximum of 42.6g in the third week of incubation. The rest of the time they are about **41g** and standard deviation around 3g.

Females arrive at 48.1g and after laying they are 50.9g. The range of mass goes from 40g up to 71g with eggs, so a reasonable 'normal' breeding season mass is **49.5g.** The standard deviation on arrival day is 4.3g.

BWP quotes the weight of eggs as 12g, (from NT2) and our few measurements are consistent with that but Glutz quotes 12.75g. They lose weight during incubation, and I suggest **12.5g** is used.

BNA quotes 9.5g and Nelson gave 9.63 from her studies in the 1930s so **9.5g** is sound. The standard deviation of the 343 eggs that Reed & Oring 1997 gave was 0.64g.

COMPARISONS CS TO SS

Male to male

On linear dimensions

Wing-length 111/105 = 1.057

Bill-length 25/24 = 1.04

Tarsus 24.2/23.8 = 1.017

On mass 50/41 = 1.22

Female to female

On linear dimensions

Wing-length 115/109 = 1.055

Bill-length 25/24 = 1.04

Tarsus 24.2/23.8 = 1.017

On mass 57/49.5 = 1.15

Egg to egg

On length	36/32 = 1.12
On breadth	26/24 = 1.08
On mass	12.5/9.5 =1.32

So the CS eggs are 30 per cent bigger in mass (consistent with egg linear dimensions being 10 per cent bigger).

The female CS is 15 per cent bigger in mass (consistent with the wing being 5 per cent bigger).

Thus we see that proportionally the SS is laying smaller eggs, as may be expected from a serial layer.

The male CS is 22 per cent bigger in mass than SS. Or to put it another way, the male CS is not quite so much smaller than its female as the spotted male is than its female This is plausibly a secondary consequence of monogamy, as the male CS is not so food-stressed during incubation; another factor is that he is incubating a relatively bigger clutch of eggs and if he was much smaller he would not cover them properly.

BIBLIOGRAPHY

Able, K.P. 1999. *Gatherings of Angels: migrating birds and their ecology.* Cornell University Press. Ithaca and London.

Adam, P. 1990. *Saltmarsh Ecology.* Cambridge University Press.

Adamik, P. & Pietruszkova, J. 2008. Advances in spring but variable autumnal trends in timing of inland wader migration. *Acta Ornithologica 43: 119–128.*

Akriotis, T. 1991. Weight changes in the Wood Sandpiper *Tringa glareola* in south-eastern Greece during the spring migration. *Ringing and Migration,* 12: 61–66.

Alberico, J.A.R., Reed, J.M and Oring L.W. 1991. Nesting near a Common Tern colony increases and decreases Spotted Sandpiper nest predation. *Auk* 108:904–910.

Alberico, J. A. R., J. M. Reed, and L. W. Oring. 1992. Non-random philopatry of sibling Spotted Sandpipers. *Ornis Scand.* 23: 504–508.

Alerstam, T. 1990. *Bird Migration*. Cambridge University Press

Altenburg, W., Engelmoer, M., Mes, R. & Piersma, T. 1982. Wintering waders on the Banc d'Arguin, Mauritania. *Communication 6 of the Wadden Sea Working Group.*

Altenberg, W. & Van der Kamp, J. 1988. Coastal Waders in Guinea. *Wader Study Group Bull.*54: 33–35.

Andersson, M., Wallander, J., Oring, L., Akst, E., Reed, J.M. & Fleischer, R.C. 2003. Adaptive seasonal trends in brood sex ratio: test in two sister species with contrasting breeding systems. *J. Evol. Biol.* 16: 510–515.

Andres, B.A., Smith, P.A., Morrison, R.I.G., Gratto-Trevor, C.L., Brown, S.C. & Friis, C.A. 2012. Population estimates of North American shorebirds, 2012. *Wader Study Group Bull.* 119(3): 178–194.

Anthes, N., Bergmann, H-H., Hegemann, A., Jaquier, S., Kriegs, J.O., Pyzhjanov, S. & Schielzeth, H. 2004. Waterbird phenology and opportunistic acceptance of a low-quality wader staging site at Lake Baikal, eastern Siberia. *Wader Study Group Bull.* 105: 75–83.

Arcas, J. 1999. Origin of Common Sandpipers *Actitis hypoleucos* captured in the Iberian peninsula during their Autumn migration. *Wader Study Group Bull.* 89: 56–59.

Arcas, J. 2000. Diet of Common Sandpiper during its Autumn Migration in the Ria de Vigo, Northwest Spain. *Alauda* 68: 265–274.

Arcas, J. 2001a. Predation of Common Sandpiper *Actitis hypoleucos* on *Orchestia gammarellus*: problems in assessing its diet from pellet and dropping analysis. *Wader Study Group Bull.* 94: 31–33.

Arcas, J. 2001b. Body weight variation and fat deposition in Common Sandpipers *Actitis hypoleucos L.* during their autumn migration in the Ria de Vigo, Galicia, north-west Spain. *Ringing and Migration* 20: 216–220.

Arcas, J. 2002. Age related differences in body mass and fat deposition of Common Sandpipers (*Actitis hypoleucos*) during their autumn migration in northwest Spain. *Alauda* 70: 323–326.

Arcas, J. 2004. Diet and prey selection of Common Sandpiper *Actitis hypoleucos* during winter. *Ardeola* 51(1): 203–213.

Ash. J.H. 1969. Spring weights of trans-Saharan migrants in Morocco. *Ibis* 111: 1–10.

Ayerbe-Quinonez, F. & Johnston-Gonzalez, R. 2010. Phenelogy of southward shorebird migration through the Popoyan Plateau, Andes of Colombia. *Wader Study Group Bull.* 117(1): 15–18.

Baccetti, N., De Faveri, A. & Serra, L. 1992. Spring migration and body condition of Common Sandpipers *Actitis hypoleucos* on a small Mediterranean island. *Ringing and Migration* 13: 90–94.

Baker, A.J., Pereira, S.L. & Paton, T.A. 2007. Phylogenetic relationships and divergence times of Charadriiformes genera: multigene evidence for the Cretaceous origin of at least 14 clades of shorebirds. *Biol. Lett* 3: 205–209.

Ballance, D.K. 2006. *A History of the Birds of Somerset.* Isabelline Books

Balmer, D.E., Gillings, S., Caffrey, B.J., Swann, R.L., Downie, I.S. & Fuller, R.J. 2013. *Bird Atlas 2007–2011; the breeding and wintering birds of Britain and Ireland.* BTO books, Thetford.

Balmori, A. 2003. Differential autumn migration of Common Sandpiper in the Duero valley (north-west Spain). *Ardeola* 50(1) : 59–66.
Balmori, A. 2005. Stopover ecology and diverse migratory strategies use in the Common Sandpiper (Actitis hypoleucos), *Ardeola* 52(2): 319–331.
Bates, B., Etheridge, B., Elkins, N., Fox, J. & Summers, R.W. 2012. Premigratory change in mass and the migration track of a Common Sandpiper *Actitis hypoleucos* from Scotland. *Wader Study Group Bull.* 119(3): 149–154.
Benson, C.W. & Irwin, M.P.S. 1974. The significance of records of the Common Sandpiper breeding in East Africa. *Bull. Br. Orn. Club*. 94:20–21.
Blackwall, J. 1822. Tables of the various species of periodical birds observed in the neighbourhood of Manchester with a few remarks tending to establish the opinion that the periodical birds migrate. *Manchester Literary and Philosophical Society Papers*
Blomqvist, D. & 9 others. 2002. Genetic similarity between mates and extra-pair parentage in three species of shorebirds. *Nature* 419: 613–615.
Boros, E., Andrikovics,S., Kiss, B. & Forro,L. 2006. Feeding ecology of migrating waders (Charadrii) at sodic-alkaline pans in the Carpathian Basin. *Bird Study* 53, 86–91.
Boyd, A.W. 1951. *A Country Parish*. Collins, London
Bradstreet,M.S.W., Page, G.W. & Johnston, W.G. 1977. Shorebirds at Long Point, Lake Erie 1966–1971, Seasonal occurrence, habitat preference and variation in abundance, *Can. Field Nat*. 91: 225–236.
Bray, J., Calladine, J & Thiel, A. 2012. Fourth year of ornithological surveys at House of Water, East Ayrshire: Breeding season 2011 and Winter season 2011–2012. *BTO Research Report No 631.*
Bregullqa, H.L. 1992. *Birds of Vanuatu*. Nelson
Brown, S.C. 1973. Common Sandpiper Biometrics. *Wader Study Group Bull.* 11:18–23.
Brucker, J.W., Gosler, A.G. & Herget, A.R. 1992. *Birds of Oxfordshire*. Pisces Publications.
Bugler, C. 2012. *The Bird in Art.* Merrell publishers, London and New York.
Burger, J. 1968. Incubation period of the Spotted Sandpiper. *Wilson Bull.* 80:101–105 and correction in 80:335.
Burkli, W. 1990. Fische als Beute des Flussuferlaufes. *Orn. Beob*. 87(1): 59.
Burton, J.F. 1995. *Birds and Climate Change.* Christoper Helm, London.
Cardoso, T.A.L., Cardoso, M.M.L., Brasilino,T. & Zeppelini, D. 2013. Distribution of shorebirds in north-eastern Brazil; preferences between open beaches and inner estuarine habitats. *Wader Study Group Bull.* 120(1): 26–31.
Clark, J.M. & Eyre, J.A. 1993. *Birds of Hampshire*, Hampshire Ornithological Society.
Cleasby, I. 1994. The 'Not-so common Sandpiper'. *Cumbrian Wildlife*. 39.
Colwell, M.A. 2010. *Shorebird Ecology, Conservation and Management*. University of California Press.
Colwell, M.A. & Oring, L.W. 1989. Extra-pair mating in the Spotted Sandpiper: a female mate acquisition tactic. *Anim. Behav*. 38: 675–684.
Coulthard, T.J., Ramirez, J.A., Barton, N., Rogerson M. & Brucher, T. 2013. Were rivers flowing across the Sahara during the last Interglacial? Implications for Human Migration through Africa. *PLoS ONE* 8(9): e74834. doi:10.1371.
Coward, T.A. & Oldham, C. 1910. *The vertebrate Fauna of Cheshire and Liverpool Bay, Vol 1 The mammals and birds of Cheshire*. London, Witherby.
Cramp, S. & Conder, P.J. 1970. A visit to the Oasis of Kufra; Spring 1969. *Ibis* 112: 261–3.
Curry-Lindahl, K. 1960. *Ecological studies on Mammals, Birds, Reptiles and Amphibians in the Eastern Belgian Congo. Part 2*. Tervuren
Cuthbertson, E.I., Foggit, G.T. & Bell, M.A. 1952. A census of Common Sandpipers in the Sedbergh area. *Brit. Birds* 45: 171–175.
D'Amico, F. 2001. Distribution morcelée et abondance du Chevalier guignette *Actitis hypoleucos* en rivière de montagne (Vallée d'Ossau, Pyrénées). *Alauda* 69(2): 223–228.
D'Amico, F. 2002. High reliability of linear censusing for Common Sandpiper (Actitis hypoleucos) breeding along upland streams in the Pyrenees, France. *Bird Study* 49: 307–309.
Davidson, N.C. 1984. How valid are flight estimates for waders? *Ringing and Migration* 5: 49–64.
Dare, P.J. 2015. *The Life of Buzzards*. Whittles Publishing, Dunbeath, Scotland: p107.
Deeming, D.C. & Reynolds, S.J. (eds.) 2015. *Nests, Eggs, & Incubation*. OUP. Oxford.
Delany, S., Scott, D., Dodman, T. & Stroud, D. (eds.) 2009. *An Atlas of Wader Populations in Africa and Western Eurasia*. Wetlands International, Wageningen, The Netherlands.
de Elgea, A.O. & Arizaga, J. 2016. Fuel load, fuel deposition rate and stopover duration of the Common Sandpiper *Actitis hypoleucos* during the autumn migration. *Bird Study* 63(2): 262–267.
del Hoyo, J., Elliott, A. & Sargatal, J. (eds.) 1996. *Handbook of the Birds of the World, Vol. 3. Hoatzin to Auks.* Lynx Edicions, Barcelona.
Dick, W.J.A. & Pienkowski, M.W. 1979. Autumn and early winter weights of waders in north-west Africa. *Ornis Scandinavica* 10: 117–123.

Didyk, A.S., Canaris, A.G. & Kinsella, J.M. 2007. Intestinal Helminths of the Spotted Sandpiper during Fall migration in New Brunswick, Canada with a Checklist of Helminths Reported from this Host. *Comp. Parasitol.* 74(2): 359–363.

Dietz, M.W., Piersma, T., Hedenstrom, A. & Brugge M 2006. Intraspecific variation in avian pectoral muscle mass: constraints on maintaining manoeuvrability with increasing body mass. *Funct.Ecol.* 21:317–326.

Diez, F. & Peris, S.J. 2001. Habitat selection by the Common Sandpiper (*Actitis hypoleucos*) in west-central Spain. *Ornis Fennica* 78: 127–134.

Dominguez, J. & Lorenzo, M. 1992. Waders wintering on the open shores of Galicia, NW Spain. *Wader Study Group Bull.* 66: 73–77.

Dougall, T.W. 2005. Hatched, matched, dispatched. *British Trust for Ornithology –RAS News* April 2005.

Dougall, T.W., Holland, P.K., & Yalden, D.W. 1995. Hatching dates for Common Sandpipers *Actitis hypoleucos* chicks – variation with place and time. *Wader Study Group Bull.* 76:53–55.

Dougall, T.W., Holland, P.K. & Yalden, D.W. 2004. A revised estimate of the breeding population of Common Sandpipers *Actitis hypoleucos* in Great Britain and Ireland. *Wader Study Group Bull.* 105: 42–49.

Dougall, T.W., Holland, P.K., & Yalden, D.W. 2010. The population biology of Common Sandpipers in Britain. *British Birds* 103: 100–114.

Dougall, T.W., Holland, P.K., Mee, A. & Yalden, D.W. 2005. Comparative population dynamics of Common Sandpipers *Actitis hypoleucos*: living at the edge. *Bird Study* 52: 80–87.

du Feu, C.R., Clark, J.A., Scaub, M., Fiedler, W. & Baillie, S.R. 2016. The EURING Data Bank – a critical tool for continental-scale studies of marked birds. *Ringing & Migration*. 31(1): 1–18.

Elas, M. & Meissner, W. 2016. Nesting phenology and breeding success of the Common Sandpiper in the Middle Vistula (Poland). *IWSG conference, Cork (poster).*

El Hamoumi, R. & Dakki, M. 2010. Phenology of waders in the Sidi Moussa-Walidia coastal wetlands, Morocco. *Wader Study Group Bull.* 117(2); 73–84.

Elkins, N. 1988. *Weather and bird behaviour, 2nd edition*. T & A D Poyser, London.

Erhart, F.C. 1997. Common Sandpiper *Actitis hypoleucos* profits from nature development projects. *Limosa* 70: 67–70.

Fefelov, I. & Tupitsyn, I. 2004. Waders of the Selenga delta, Lake Baikal, eastern Siberia. *Wader Study Group Bull.* 104: 66–78.

Ferguson, D. 2012. *The Birds of Buckinghamshire 2nd edition.* Bucks Bird Club.

Ferguson-Lees, J., Castle, P. & Cranswick, P. 2007. *Birds of Wiltshire*. Wiltshire Ornithological Society.

Fisher, J. 1966. *The Shell Bird Book.* Ebury Press and Michael Joseph

Fivizzani, A. J. & Oring, L.W. 1986. Plasma steroid hormones in relation to behavioral sex role reversal in the Spotted Sandpiper, *Actitis macularia*. *Biol. Reprod*. 35:1195–1201.

Francis, I.S., Gartshore, M, Green,M., Penford, N. & Ryall, C. 1994. Some observations of waders and other wetland birds from Ivory Coast. *Wader Study Group Bull.*72: 22–24.

Fray, R., Davis R., Gamble, D., Harrop, A. & Lister, S. 2009. *Birds of Leicestershire and Rutland,* Christopher Helm, London

Gauthier, J. & Aubry, Y. 1996. *The Breeding Birds of Quebec.* Canadian Wildlife Service, Environment Canada.

Gbogbo, F., Adaworomah, B.B., Asante, E & Brown-Engman, G.R. 2013. Human related bird flushes are of little consequence to wintering waterbirds in a tropical coastal wetland in Ghana. *Wader Study Group Bull.* 120(1): 60–65.

Geering, A., Agnew, L. & Harding, S. 2007. *Shorebirds of Australia*. CSIRO publishing.

Ghaemi, R. 2006. Shorebirds of the wetland of Gomishan, Iran. *Wader Study Group Bull.* 109: 102–105.

Gibbons, D.W., Reid, J.B. & Chapman, R.A. 1993. *The New Atlas of Breeding Birds in Britain and Ireland: 1988–1991.* Poyser, London.

Gladwin, T. & Sage, B.L. 1986 *The Birds of Hertfordshire.* Castlemead publications.

Glutz von Blotzheim, U.N., Bauer, K.M. & Bezzel, E. 1977. *Handbuch der Vögel Mitteleuropas. Vol. 7.* Aula-Verlag, Wiesbaden.

Goodfellow, P. 1983. *Shakespeare's Birds*. Penguin Books, England.

Goss-Custard, J.D.(ed.). 1996. *The Oystercatcher: from individuals to populations*. Oxford Univ. Press.

Hagemeijer, E.J.M. & Blair, M.J. (Eds). 1997. *The EBCC Atlas of European Birds: Their Distribution and Abundance.* T & A D Poyser, London.

Hale, W.G. 1980. *Waders.* Collins New Naturalist, London.

Harrison G. & Harrison J. 2005. *The New Birds of the West Midlands*. West Midlands Bird Club.

Hassell, C. 2006. Common Sandpiper feeding on Hippopotamus? *Wader Study Group Bull.* 109: 124.

Hays, H. 1972. Polyandry in the Spotted Sandpiper. *The Living Bird* 11: 43–57.

Hillman, J.C., Largen, M.J. & Yalden, D.W. 1986. Observations of a migrant Common Sandpiper in Ethiopia. *Wader Study Group Bull.* 48: 17.

Holland, P.K. 2009. Relationship between Common Sandpipers *Actitis hypoleucos* breeding along the River Lune,

England, and those fattening for migration near its mouth with a model of their onward migration. *Wader Study Group Bull.* 116(2): 83–85.
Holland, P.K., Robson, J.E. & Yalden, D.W. 1982a. The breeding biology of the Common Sandpiper *Actitis hypoleucos* in the Peak District. *Bird Study* 29: 99–110.
Holland, P.K., Robson, J.E. & Yalden, D.W. 1982b. The status and distribution of the Common Sandpiper (Actitis hypoleucos) in the Peak District. *Naturalist* 107: 77–86.
Holland, P.K. & Yalden, D.W. 1991a. Growth of Common Sandpiper chicks. *Wader Study Group Bull.* 62: 13–15.
Holland, P.K. & Yalden, D.W. 1991b. Population dynamics of Common Sandpipers Actitis hypoleucos breeding along an upland river system. *Bird Study* 38: 151–159.
Holland, P.K. & Yalden, D.W. 1994. An estimate of lifetime reproductive success for the Common Sandpiper *Actitis hypoleucos*. *Bird Study* 41: 110–119.
Holland, P.K. & Yalden, D.W. 1995. Who lives and who dies? The impact of severe April weather on breeding Common Sandpipers *Actitis hypoleucos. Ringing & Migration* 16: 121–123.
Holland, P.K. & Yalden, D.W. 2002. Population dynamics of Common Sandpipers *Actitis hypoleucos* in the Peak District of Derbyshire – a different decade. *Bird Study* 49: 131–138.
Holland, P.K. & Yalden, D.W. 2012. Observations on desertion and recruitment in a population of breeding Common Sandpipers *Actitis hypoleucos* at the edge of their range. *Wader Study Group Bull.* 119(3): 172–177.
Hollom, P.A.D., Porter, R.F., Christensen, S. & Willis, I. 1988. *Birds of the Middle East and North Africa.* T & A D Poyser, Calton.
Holloway, S. 1996. *The Historical Atlas of Breeding Birds in Britain and Ireland: 1875–1900.* T & A.D. Poyser, London .
Hotker, H., Lebedeva, E., Tomkovitch, P.S., Gromadzka, J. Davidson, N.C., Evans, J., Stroud, D.A. & West R.B. (eds) 1998. *Migration and international conservation of waders. Research and conservation on north Asian, African and European flyways.* International Wader Series 10.
Hoye, B.J. & Buttermer, W.A. 2011. Inexplicable inefficiency of avian molt? Insights from an opportunistically breeding arid-zone species, *PLoS ONE* 6(2): e16230.
Ingram, G.C.S. 1945. Common Sandpipers on migration in South Glamorgan. *British Birds* 38:169–172.
Iwajomo, S.B. & Hedenstrom, A. 2011. Migration patterns and morphometrics of Common Sandpipers at Ottenby, southeastern Sweden. *Ringing & Migration* 26: 38–47.
James, P. 1996. *The Birds of Sussex.* Sussex Ornithological Society.
Jenkins, D. & Sparks, T.H. 2010. The changing bird phenology of Mid Deeside, Scotland 1974–2010. *Bird Study* 57: 407–414.
Johns, C.A. 1910, *British Birds in their Haunts.* Routledge, London.
Johnston-Gonzales, R., Castillo, L.F., Hernandez, C & Ruiz-Guerra, C. 2006. Whimbrels roosting in Colombian mangroves. *Wader Study Group Bull.* 110: 63
Kajtoch, L. & Figarski, T. 2013. Short-term restoration of riverine bird assemblages after a severe flood. *Bird Study* 60: 327–334.
Kirk, G. & Phillips, J. 2013. *The Birds of Gloucestershire.* Liverpool University Press.
Kitson, A. 1979. Notes on the Waders of Mongolia. *Wader Study Group Bull.* 27: 34–35.
Kruckenberg, H., Kondratyev, A., Zockler, C., Zaynagutdinova, E. & Mooij, J.H. 2012. Breeding Waders on Kolguev Island, Barents Sea, N. Russia, 2006–8. *Wader Study Group Bull.* 119(2) 102–113.
Kuenzel, W. J. & Wiegert, R.G. 1973. Energetics of a Spotted Sandpiper feeding on brine fly larvae (*Paracoenia; Diptera; Ephydridae*) in a thermal spring. *The Wilson Bulletin.* 85, No.4: 479–476.
Kuwae, T., Beninger, P.G., Decottignies, P., Mathot, K.J., Lund, D.R. & Elner, R.W. 2008. Biofilm grazing in a higher vertebrate: the Western Sandpiper, Calidris mauri. *Ecology* 89(3) 599–606.
Lack, D. 1968. *Ecological adaptations for Breeding in Birds.* Methuen: London
Lack, P. 1986. *The Atlas of wintering birds in Britain and Ireland.* T & A D Poyser.
Lane, B.A. 1987. *Shorebirds in Australia.* Nelson, Melbourne.
Lank, D. B., Oring, L.W. & Maxson, S.J. 1985. Mate and nutrient limitation of egg-laying in a polyandrous shorebird. *Ecology* 66:1513–1524.
Lappo, E., Tomkovitch, P. & Syroechkovskiy. 2012. *Atlas of breeding waders in the Russian Arctic.* Moscow.
Lawrence, D. 1993. Spotted Sandpiper displaying to and mating with Common Sandpiper. *Br. Birds* 86:628.
Lecoq, M., Lourenco, P.M., Catry, P., Andrade, J & Granadeiro, J.P. 2013. Wintering waders on the Portuguese mainland non-estuarine coast: results of the 2009–2011 survey. *Wader Study Group Bull.* 120(1): 66–70.
Lofaldi, L. 1981. On the breeding season biometrics of the Common Sandpiper. *Ringing & Migration* 3(3): 133–136.
Lunn, J. 1992. One hundred years on –a comparison of the arrival dates of summer migrants in the Barnsley area. *Magpie* 4: 34–40.
Lyngs, P. 1996. Waterbirds at Lake Oloidien, Naivasha, Kenya, autumn 1987. *Wader Study Group Bull.* 79: 91–102.
Marchant, J.H., Hudson, R., Carter, S.P. & Whittington, P. 1990. *Population Trends in British Breeding Birds.* BTO Tring.

Mason, C.F. 1984. The passage of waders at an inland reservoir in Leicestershire. *Ringing & Migration* 5: 133–140.
Matthiessen, P. 1981. *Sand Rivers*. Bantam Books: p35.
Maxson, S.J. & Oring, L.W. 1978. Mice as a Source of Egg Loss among Ground-nesting Birds. *Auk* 95: 582–584.
Maxson, S.J. & Oring, L.W. 1980. Breeding season time and energy budgets of the polyandrous Spotted Sandpiper. *Behaviour* 74:200–263.
Mcneil, R. 1970. Hivernage et estivage d'oiseaux aquatiques nord-américains dans le Nord-Est du Venezuela (mue, accumulation de graisse, capacité de vol et routes de migration). *Oiseau* 40:185–302.
McNicholl, M.K. 1981. Egg-teeth of Spotted Sandpipers. *North American Bird Bander* 6(2): 44–45.
Mee, A. 2001. *Reproductive Strategies in the Common Sandpiper.* PhD thesis University of Sheffield.
Mee, A., Whitfield, D.P., Thompson, D.B.A. & Burke, T. 2004. Extrapair paternity in the Common Sandpiper *Actitis hypoleucos. Animal Behaviour* 67: 333–342.
Meissner, W. 1996. Timing and phenology of autumn migration of Common Sandpiper (*Actitis hypoleucos*) at the Gulf of Gdansk. *The Ring* 18,1–2: 59–72.
Meissner, W. 1997. Autumn migration and biometrics of the Common Sandpiper *Actitis hypoleucos* caught in the Gulf of Gdansk. *Ornis Fennica* 74: 131–139.
Meissner, W., Holland, P.K. & Cofta, T. 2015. Ageing and sexing the Common Sandpiper *Actitis hypoleucos. Wader Study* 122(1): 54–59.
Mester, H. 1966. Zuggewohnheiten sowie Grossen- und Gewichts-Variationen des Flussuferlaufers (*Tringa hypoleucos*). *Die Vogelwarte* 23.4: 291–300.
Miller, J. R. & Miller, J.T. 1948. Nesting of the Spotted Sandpiper at Detroit, Michigan. *The Auk* 65: 558–567.
Milligan, B.S. 1979. Some observations of wader passage on the south Kenya coast during 1978. *Wader Study Group Bull.* 25: 28–30.
Mitchell, F.S. 1885. *The Birds of Lancashire*. London: Van Voorst.
Moggridge, M. 1851. The Kestril in pursuit of prey. *Ann. Mag. Nat. Hist.* (2)7: 501.
Montier, D. 1977. *Atlas of Breeding Birds of London Area*. Batsford Ltd.
Mortensen, S. & Ringle, J. 2007. Changes in Colonial Waterbird Populations on Leech Lake. *The Loon* 79(3): 130–142.
Mousley, H. 1939. Nesting behaviour of Wilsons Snipe and Spotted Sandpiper. *The Auk* 56:129–133.
Naves, L.C., Lanctot, R.B., Taylor, A.R. & Coutsobos, N.P. 2008. How often do Arctic shorebirds lay replacement clutches? *Wader Study Group Bull.* 15(1):2–9
Nelson, T. 1930. Growth rate of the Spotted Sandpiper chick with notes on nesting habit. *Bird Banding*, 1:1–13.
Nethersole-Thompson, D. 1951 *The Greenshank* Collins.
Nethersole-Thompson, D. & Nethersole-Thompson, M. 1986. *Waders their breeding, haunts and watchers.* T & A.D. Poyser, Calton.
Newton, I. 1986. *The Sparrowhawk*. T & A D Poyser Ltd. Calton.
Newton, I. 2008. *The Migration Ecology of Birds.* Academic Press.
Nicoll, M. & Kemp, P. 1983. Partial primary moult in first-spring/summer Common Sandpipers *Actitis hypoleucos. Wader Study Group Bull.* 37: 37–38.
Nisbet, I.C.T. 1957. Wader migration at Cambridge Sewage-farms. *Bird Study* 4: 131–148.
Ntiamoa-Baidu, Y. & Grieve, A. 1987. Palearctic waders in coastal Ghana in 1985/6. *Wader Study Group Bull.* 49 supplement (IWRB special pub 7): 76–78.
OAG Munster 1982. Inland Wader Counts – Second Progress Report. *Wader Study Group Bull.* 35: 11–13.
Ogilvie, M. 1994. *British Birds* 87: 384.
Oring, L.W. & Knudson, M.L. 1972. Monogamy and polyandry in the Spotted Sandpiper. *Living Bird* 11:59–73.
Oring, L.W. & Lank, D.B. 1982. Sexual Selection, Arrival Times, Philopatry and Site Fidelity in the Polyandrous Spotted Sandpiper. *Behav. Ecol. Sociobiol.* 10: 185–191.
Oring, L.W. & Maxson, S.J. 1978. Instances of simultaneous polyandry by a Spotted Sandpiper *Actitis macularia. Ibis* 120: 349–353.
Oring, L.W., Reed, J.M., Colwell, M.A., Lank, D.B. & Maxson, S.J. 1991. Factors regulating annual mating success and reproductive success in Spotted Sandpipers (*Actitis macalauria*). *Behav. Ecol. Sociobiol.* 28: 433–442.
Oring, L.W., Fleischer, R.C., Reed, J.M. & Marsden, K.E. 1992. Cuckoldry through stored sperm in the sequentially polyandrous Spotted Sandpiper. *Nature* 359: 631- 633.
Oring, L. W., Fivizzani, A.J. & El Halawani, M.E. 1986a. Changes in plasma prolactin associated with laying and hatch in the Spotted Sandpiper. *Auk* 103:820–822.
Oring, L. W., A. J. Fivizzani, M. E. El Halawani, and A. Goldsmith. 1986b. Seasonal changes in prolactin and luteinizing hormone in the polyandrous Spotted Sandpiper, *Actitis macularia*. *Gen. Comp. Endocrin.* 62:394–403.
Oring, L. W., A. J. Fivizzani, M. E. El Halawani 1989 Testerone-induced inhibition of incubation in the Spotted Sandpiper (Actitis macalauria) *Hormones and behaviour* 23, Iss 5 412–423.
Oring, L. W., J. M. Reed, and S. J. Maxson. 1994b. Copulation patterns and mate guarding in the sex-role reversed, polyandrous Spotted Sandpiper, *Actitis macularia. Anim. Behav.* 47:1065–1072.

Parmelee, D.F. & Payne, R.B. 1973. On multiple Broods and the breeding strategy of Arctic Sanderlings. *Ibis* 115: 218–226.

Pearce-Higgins, J.W., Yalden, D.W., Dougall, T.W., Beale, C.M. 2009. Does climate change explain the decline of a trans-Saharan Afro-Palearctic migrant? *Oecologia* 159: 649–659.

Pearson, D.J. 1974. The timing of wing moult in some Palaearctic waders wintering in East Africa. *Wader Study Group Bull*. 12: 10–17.

Pearson, D.J. 1977. The first year moult of the Common Sandpiper in Kenya. *Scopus* 1(4): 89–94.

Penhallurick, R.D. 1969. *The Birds of the Cornish Coast*. D Bradford Burton Ltd.

Pennycuick, C.J. 2008. *Modelling the flying bird*. Elsevier, Amsterdam

Perco, F. 1984. Estimates of wader numbers during midwinter in Northern Adriatic coastal wetlands. *Wader Study Group Bull*. 40: 49–50.

Perkins, G.A. & Lawrence J.S. 1985. Bird use of wetlands created by surface mining. *Trans. Illinois State Acad. Sci.* 78: 87–96.

Pickett, P.E., Maxson S.J. & Oring L.W. 1988. Interspecific interactions of Spotted Sandpipers. *Wilson Bull.* 100:297–302.

Pickett, P.E., Oring, L.W. & Fivizzani, A.J. Jr. 1989. First documented case of a captive-reared sandpiper breeding in the wild. *J. Field. Ornithol*. 60(3): 312–314.

Pienkowski. M.W., Knight, P.J., Stanyard, D.J. & Argy;e, F.B. 1976, The Primary Moult of Waders on the Atlantic Coast of Morocco. *Ibis* 118: 347–365

Piersma, T. & van Gils, J.A. 2011. *The Flexible Phenotype*. Oxford University Press.

Prater, A.J., Marchant, J.H. & Vuorinen, J. 1977. *Guide to the identification and ageing of Holarctic waders*. BTO Guide 17.

Preston, F.W. 1951. Egg-laying, incubation and fledging periods of the Spotted Sandpiper. *Wilson Bull*. 63:43–44.

Reed, J.M. & Oring, L.W. 1992. Reconnaissance for future breeding sites by Spotted Sandpipers. *Behav Ecol* 3:310–317.

Reed, J. M. & Oring, L.W. 1993. Philopatry, site fidelity, dispersal, and survival of Spotted Sandpipers. *The Auk* 110: 541–551

Reed, J. M. & Oring, L.W. 1997. Intra- and inter-clutch patterns in egg mass in the Spotted Sandpiper. *J. Field Ornithol.* 68(2): 296–301.

Reed, J. M., Fleischer, R.C., Eberhard, J & Oring, L.W. 1996. Minisatellite DNA variability in two populations of Spotted Sandpipers *Actitis macularia* in Minnesota, U.S.A. *Wader Study Group Bull.* 79:115–117.

Remisiewicz, M. 2011. The flexibility of primary moult in relation to migration in Palaearctic waders – an overview. *Wader Study Group Bull.* 118(3): 163–174.

Riddiford, N. & Findley, P. 1981. *Seasonal Movements of Summer Migrants*. British Trust for Ornithology- guide 18.

Riedel, B. 1978. Zur Zugbiologie des Flussuferlaufers auf den Nortener Schlammteichen. *Faun. Mitt. Sud-Niedersachsen* 1: 199–213.

Robinson, R.A. and 14 others 2015. Birdtrends 2015. Research Report 678. BTO Thetford (on-line)

Roche, J. 1989. Distribution du Chevalier Guignette et de L'ombre Commun de long des rivieres de France et d'Europe. *Bull. Ecol.* 20,3:231–236.

Round, P.D. & Moss,M. 1984. The waterbird population of three Welsh Rivers. *Bird Study* 31: 61–68.

Round, P.D., Gale, G.A. & Nimnuan, S. 2012. Moult of primaries in Long-toed Stints at a non-breeding area in Thailand. *Ringing & Migration* 27 Part 1: 32–37.

Rufino, R., Neves, R. & Pina, J.P. 1998.Wintering waders in Dakhla Bay, Western Sahara. *Wader Study Group Bull.*87: 26–29.

Sage, B.L. 1969. *A History of the Birds of Hertfordshire*. Barrie & Rockliff, London.

Sandercock, B.K., Casey, A., Green, D.E., Ip, H.S. & Converse, K.A. 2008. Reovirus associated with mortality of an Upland Sandpiper. *Wader Study Group Bull.* 115(1): 55–56.

Sandilyan, S., Thiyagesan, K. & Nagarajan, R. 2010. Major decline in species-richness of waterbirds in the Pichavaram mangrove wetlands, southern India. *Wader Study Group Bull*. 117(2): 91–98.

Sauer, J.R., Niven, D.K., Hines, J.E., Ziolowski, D.J., Pardiek, K.L., Fallon, J.E. & Link, W.A. 2017, The North American BBS results 1966–2015. *USGS Patuxent Wildlife Research Centre* (online).

Sauvage, A., Rumsey, S. & Rodwell, S. 1998. Recurrence of Palaearctic birds in the lower Senegal river valley. *Malimbus* 20: 33- 45.

Schekkerman, H. & Visser, G.H. 2001. Prefledging energy requirements in shorebirds: energetic implications of self-feeding precocial development. *The Auk* 118(4): 944–957.

Schmitz, M., Sudfeldt, C., Legge, H., Mantel, K.,Weber, P. & Marinov, M. 1999. Spring migration of waders in the Razim-Sinoie lagoon system south of the Danube delta, Romania. *Wader Study Group Bull*. 90: 59–64.

Serasinghe, R. 1992. Communal roosting of the Common Sandpiper (*Actitis hypoleucos*). *Loris* 19(5):174–175.

Sharrock, J.T.R. 1976. *The Atlas of Breeding Birds in Britain and Ireland*, British Trust for Ornithology and Irish Wildbird Conservancy

Simmons, K.E.L. 1951. Behaviour of Common Sandpiper in winter quarters. *British Birds* 44: 415–416.
Sitters, H.P. 1988. *Tetrad Atlas of the Breeding Birds of Devon*. DBWPS
Smit, C. 1986 Wintering and migrating waders in the Mediterranean *Wader Study Group Bull*. 46: 13–15.
Soares, L., Escudero, G., Peulia, V.A.S. & Ricklefs, R.E. 2016. Low prevalence of haemosporidian parasites in shorebirds, *Ardea*.104: 129–141.
Soikkeli, M. 1967. Breeding cycle and population dynamics in the dunlin (*Calidris alpina*). *Annales Zoologica Fennica* 4:158–198.
Spaans, A.L. 1978. Status and numerical fluctuations of some North American waders along the Surinam coast. *Wilson Bull.* 90:60–83.
Spaans, A.L. 1979. Wader studies in Surinam, South America *Wader Study Group Bull*. 25: 32–37.
Spalding, M., Kainuma, M. & Collins, L. 2010. *World Atlas of Mangroves*. Earthscan.
Spiekman, H. 1992. Results of wader ringing in Tunisia 1962–1986. *WIWO-report* nr.44.
Staaland, H. 1967. Anatomical and Physiological adaptations of the Nasal Glands in Charadriiformes. *Comp. Biochem. Physiol.* 23: 933–944.
Standley, P., Bucknell N.J., Swash, A. & Collins I.D. 1996. *Birds of Berkshire*, Berkshire Atlas Group.
Stanton, D.J. 2013a. Common Sandpiper *Actitis hypoleucos* feeding on a gecko. *Wader Study Group Bull*. 120(3): 210
Stanton, D.J. 2013b. Mobbing of a snake by a non-breeding Common Sandpiper *Actitis hypoleucos. Wader Study Group Bull.* 120(3): 210
Straw, P. 2005. Status and Conservation of Shorebirds in the East Asian-Australasian Flyway; Proceedings of the Australasian Shorebirds Conference 2003, Canbera. Wetlands International Global Series 18, International Wader Series 17. Sydney, Australia.
Sultana. J. & Gauci, C. 1982. *A new guide to the birds of Malta*. The Ornithological Society. Malta.
Summers, R.W., Underhill, L.G., Pearson, D.J. & Scott D.A. 1987 Wader migration systems in southern and eastern Africa and western Asia. WSGB 49 supplement (IWRB special publication 7): 15–34
Swennen, K & Saeho, S. 1993. Eliciting probing in a non-probing wader species, the Common Sandpiper. *Wader Study Group Bull.* 70: 31–32.
Tapper, S. 1992. *Game Heritage – An Ecological Review from Shooting and Gamekeeping Records.* Game Conservancy Ltd, Fordingbridge
Tatner, P. & Bryant, D.M. 1993. Interspecific variation in daily energy expenditure during avian incubation. *J. Zool., Lond* 231:215–232.
Taylor, M., Seago, M., Allard, P. & Dorling, D. 1999. *The Birds of Norfolk*.Helm
Ticehurst, C.B. 1932. *A History of the Birds of Suffolk*. Oliver and Boyd.
Tree, A.J. 1966. Further records of Palaeactic birds returning to the place of banding in Zambia. *Ostrich* 37: 196.
Tree, A.J. 1974. The use of primary moult in ageing the 6–15 month age class of some Palaearctic waders. *Safring News* 3(3): 21–24
Tree, A.J. 2008. The Common Sandpiper in Zimbabwe. *Honeyguide 54* (1&2): 40–51.
Trodd, P. & Kramer, D. 1991. The *Birds of Bedfordshire*, Castlemead publications
Trolliet, B.& Fouquet, M. 2004. Wintering waders in coastal Guinea. *Wader Study Group Bull*. 103: 56–62
Tye, A. & Tye, H. 1987. The importance of Sierra Leone for wintering waders. *Wader Study Group Bull.* 49 supplement (IWRB special pub 7):71–75.
Uglow, J. 2006. *Nature's Engraver; A Life of Thomas Bewick.* Faber and Faber.
Underhill, L.G. 1994. Reviving the calcium-from-Lemmings hypothesis. *Wader Study Group Bull.* 75: 35–36.
Underhill L 1999 Review of ring recoveries of waterbirds in southern Africa
Van der Winden, J., Siaka, A., Dirksen, S. & Poot, M.J.M. 2009. New estimates for wintering waders in coastal Sierra Leone. *Wader Study Group Bull.* 116(1): 29–34
Vansteenwegen, C. 1978. Le Chevalier guignette et d'autre limicoles en halte de migration autumnal. *Aves* 15: 86–122.
Velasco, T. 1992. Waders along inland rivers in Spain. *Wader Study Group Bull.* 64: 41–44.
Vernon, C.J. 1965. Incidents of aggressive behaviour by the willow warbler in its winter quarters. *Bull. B.O.C.* 85:152.
Vickery, J. 1991. Breeding density of Dippers, Grey Wagtails and Common Sandpipers in relation to the acidity of streams in south-west Scotland. *Ibis* 133: 178–185.
Viksne, J.A. & Michelson, H.A. 1985, *Migrations of birds of Eastern Europe and Northern Asia.* Nauka, Moscow.
Voous, K.H. 1960. *Atlas of European Birds*, Nelson, London.
Ward. R.M. 1999. The shorebirds of Gharo Creek and the Indus Delta, Pakistan. *Wader Study Group Bull.* 90: 31–34.
Wasser, S.K. 1983. *Social Behaviour of Female Vertebrates*. Academic Press.
Watson, A., Nethersole-Thomson, D., Duncan, K., Galbraith, H., Rae, S., Smith, R. & Thomas, C. 1988. Decline of shore waders at Loch Morlich. *Scott. Birds* 15: 91–92.

Wetmore, A. 1916. Birds of Porto Rico. *Washington DC, Dept of Agriculture Bulletin 326 .*

Wheatley, J.J. 2007. *Birds of Surrey*, Surrey Bird Club.

White, C.M.N. 1975. Migration of Palaearctic Waders in Wallacea. *Emu* 75: 37–39.

White, G. 1768. *The Natural History of Selborne*. LetterXX to Pennant.

Whitlock, F.B. 1893. *The Birds of Derbyshire*. London

Whittaker, I. 1932. *The Birds of Heywood district*. Robert Howe Ltd, Heywood (in Bury library).

Wilson, G.E. 1976. Spotted Sandpiper nesting in Scotland. *British Birds* 69: 288–292.

Wilson, J.D. 1985. *Birds and birdwatching at Pennington Flash*. Pennington Flash Committee

Wood, C. 2012. Patch feeding by Common Sandpipers *Actitis hypoleucos. Wader Study Group Bull*. 119(2): 142–143.

Wood, S. 2007. *The Birds of Essex*, Christopher Helm, London.

Yalden, D.W. 1984. Common Sandpiper numbers and recreational pressures in the Derwent Valley. *Magpie* 3: 38–46.

Yalden, D.W. 1986a. Diet, food availability and habitat selection of breeding Common Sandpiper *Actitis hypoleucos. Ibis* 128: 23–36.

Yalden, D.W. 1986b. The habitat and activity of Common Sandpiper *Actitis hypoleucos* breeding by upland rivers. *Bird Study* 33: 214–222.

Yalden, D.W. 1992a. The influence of recreational disturbance on Common Sandpipers *Actitis hypoleucos* breeding by an upland reservoir in England. *Biological Conservation* 61: 41–49.

Yalden, D.W. 1992b. The Common Sandpiper population of the Ladybower reservoir complex – a re-evaluation. *Naturalist* 117: 63–68.

Yalden, D.W. 1999. *The History of British Mammals*. T & A D Poyser Ltd. London.

Yalden, D.W. 2012. Wing area, wing growth and wing loading of Common Sandpipers *Actitis hypoleucs*. *Wader Study Group Bull.* 119(2): 84–88.

Yalden, D.W. & Albarella, U. 2009. *The History of British Birds*. OUP Oxford.

Yalden, D.W. & Dougall, T.W. 1994. Habitat, Weather, and the Growth Rates of Common Sandpiper Chicks. *Wader Study Group Bull.* 73: 33–35.

Yalden, D.W. & Dougall, T.W. 2003. Production, survival and catchability of chicks of Common Sandpipers. *Wader Study Group Bull.* 104: 82–84.

Yalden, D.W. & Holland P.K. 1992. Relative contributions of Common Sandpiper *Actitis hypoleucos* parents to guarding their chicks. *Ringing and Migration* 13: 95–97.

Yalden, D.W. & Holland P.K. 1993. Census-efficiency for breeding Common Sandpipers *Actitis hypoleucos*. *Wader Study Group Bull.* 71: 35–38.

Yalden, P.E & Yalden, D.W. 1990. Recreational disturbance of breeding golden plovers *Pluvialis apricarius Biol Conserv* 51(4): 243–262.

Ydenberg, R.C., Butler, R.W., Lank, D.B., Smith, B.D. & Ireland, J. 2004. Western Sandpipers have altered migration tactics as peregrine falcon populations have recovered. *Proc. Royal Soc: Series B:* Vol. 271 Issue 1545: 1263–1269.

Zink, R.M., Pavlova, A., Drovetski, S. & Rohwer, S. 2008. Mitochondrial phylogeographies of five widespread Eurasian bird species. *J.Ornithol* 149:399–413.

Zwarts, L. 1985. The winter exploitation of Fiddler Crabs *Uca tangeri* by waders in Guinea-Bissau. *Ardea* 73: 3–12.

Zwarts, L., 1988. Numbers and distribution of coastal waders in Guinea-Bissau. *Ardea* 76: 42–55.

Zwarts, L., Felemban, H. & Price, A.R.G. 1991. Wader counts along the Saudi Arabian coast suggests the Gulf harbours millions of waders, *Wader Study Group Bull.* 63: 25–32.

Zwarts L., Bijlsma R.G., van der Kamp J. & Wymenga E. 2009. *Living on the edge: Wetlands and birds in a changing Sahel*. KNNV Publishing, Zeist, The Netherlands.

INDEX